Clean Energy Technologies - Hydrogen and Gasification Processes

Edited by Murat Eyvaz, Yongseung Yun and Ahmed Albahnasawi

Published in London, United Kingdom

Clean Energy Technologies - Hydrogen and Gasification Processes
http://dx.doi.org/10.5772/intechopen.102298
Edited by Murat Eyvaz, Yongseung Yun and Ahmed Albahnasawi

Contributors
Indra Nath Chakraborty, Sibel Irmak, Changqing Cao, Lihui Yu, Wenhao Li, Lanjun Liu, Peigao Duan, Shehzaad Kauchali, Shaakirah Cassim, Akhoury Sudhir Kumar Sinha, Umaprasana Ojha, Zahoor Alam, Ajay Awdheshprasad Tripathi, Francisco Brójo, Rafael Domingues, Esin Iplik, David Muren, Martin Adendorff, Vikram Rama Uttam Pandit, Shivaji Tulsa Sandu Pandit, Ganesh Punjaba Dawange, Aeed S. S. Al-Fahdawi, Nurlan Bakranov, Ganesh Kavita Parshuram Jadhav, Omkar Sadhna Arun Malusare, Murat Eyvaz, Ahmed Albahnasawi

First published in London, United Kingdom, 2022 by IntechOpen
IntechOpen is the global imprint of INTECHOPEN LIMITED, registered in England and Wales, registration number: 11086078, 5 Princes Gate Court, London, SW7 2QJ, United Kingdom

British Library Cataloguing-in-Publication Data
A catalogue record for this book is available from the British Library

Additional hard and PDF copies can be obtained from orders@intechopen.com

Clean Energy Technologies - Hydrogen and Gasification Processes
Edited by Murat Eyvaz, Yongseung Yun and Ahmed Albahnasawi
p. cm.
Print ISBN 978-1-80355-402-0
Online ISBN 978-1-80355-403-7
eBook (PDF) ISBN 978-1-80355-404-4

Meet the editors

Dr. Murat Eyvaz is an associate professor in the Environmental Engineering Department, at Gebze Technical University, Turkey. His research interests include applications in water and wastewater treatment facilities, electrochemical treatment processes, filtration systems at the lab and pilot scale, membrane processes (forward osmosis, reverse osmosis, membrane bioreactors), membrane manufacturing methods (polymeric membranes, nanofiber membranes, electrospinning), spectrophotometric analyses (UV, atomic absorption spectrophotometry), and chromatographic analyses (gas chromatography, high-pressure liquid chromatography). He has co-authored many journal articles and conference papers and has taken part in several national projects. He serves as an editor and reviewer for numerous journals. He has four patents on wastewater treatment systems.

Dr. Yongseung Yun received his Ph.D. in Chemical Engineering from the University of Utah, USA, in 1990. He obtained his MA from the Korea Advanced Institute of Science and Technology (KAIST) in 1981, and his BA from Yonsei University, Korea, in 1979. He is currently a vice president at the Institute for Advanced Engineering, Korea. He has been working on gasification technology development since 1994, including coal gasification, municipal solid waste gasification, and petroleum coke gasification. He currently heads the 25-ton/day gasification project in Korea to produce blue hydrogen. He was the president of the Korea Association of Waste to Energy Technology (KAWET) from 2013 to 2019 and has been the vice president of the Korea Dimethyl ether (DME) Association since 2008. Dr. Yun was the editor of the *Korean Industrial Chemistry News* from 2009 to 2016.

Dr. Ahmed Albahnasawi is a research fellow in the Environmental Engineering Department, at Gebze Technical University, Turkey. His graduate work focused on the investigation of the treatability of the sequential anoxic-aerobic batch reactors followed by ceramic membrane for textile wastewater treatment. Based on his Ph.D. research, Dr. Albahnasawi has published numerous journal articles and participated in international conferences. His research interests include the application and design of microbial fuel cells integrated with Fenton oxidation for industrial wastewater treatment/solid waste management and monitoring of organic micropollutants by both chromatographic and spectrophotometric analyses.

Contents

Preface

Clean energy is defined as energy obtained from renewable and zero-emission sources; it also refers to a range of environmentally friendly energy options derived from renewable, low-emission sources such as solar geothermal energy, hydropower, ocean power, and wind power. Unlike fossil energy sources, clean energy sources do not run out and constantly renew themselves. For this reason, clean energy is also called sustainable energy. Other natural resources such as plants and animal wastes and hydrogen are also among the elements from which clean energy is obtained. Unlike depleting fossil energy sources, clean energy is less costly. It does not harm the environment, contributes to the protection of nature, and is the most efficient use of nature. Today, countries adopt clean energy technologies and infrastructure, invest in renewable energy sources, and prioritize energy efficiency practices to accelerate the transition to an affordable, reliable, and sustainable energy system.

Clean energy is one of the most effective ways to combat climate change. While clean energy can be defined as energy obtained from sources that do not pollute the air, produce greenhouse gas emissions, or cause any harm to nature, green energy is the energy obtained from natural sources. On the other hand, renewable energy is produced from continuously renewable and inexhaustible sources. The direct use of renewable energy in reducing carbon emissions is more on the agenda today. Effective improvements in energy efficiency, the importance of which is often overlooked, are also considered. Issues such as the greater inclusion of electric vehicles in the transportation system, the widespread use of heat pumps in residences, the inclusion of clean hydrogen and its derivatives in the energy system, and the use of bioenergy, carbon capture, and storage technologies come to the fore.

Hydrogen, one of the clean energies, has the potential to solve the growing energy crisis today due to its clean, renewable, high-energy density, and non-carbon fuel properties. Hydrogen is a synthetic fuel that can be produced from various raw materials such as water, fossil fuels, and biomass using primary energy sources. During the production phase, there are many alternative production technologies such as steam recovery, waste gas purification, electrolysis, photo processes, thermochemical processes, and radiolysis. Hydrogen has a variety of uses, including nonpolluting vehicles, fuel cells, home heating systems, and aircraft. Furthermore, using hydrogen as an energy carrier is a long-term option for reducing global carbon dioxide emissions by obtaining high-value hydrocarbons through carbon dioxide hydrogenation. The costs of a hydrogen-based system, however, are still high. Therefore, research on hydrogen, which is abundant in nature, continues so that hydrogen-based systems, particularly for long-term energy storage, can become commercially attractive. The use of hydrogen energy as a complement to traditional energy sources also aids in the implementation of low-carbon solutions.

This edited collection consists of two sections and nine chapters. The first section includes case studies for hydrogen production research. The second section presents research on gasification, a process in which a limited amount of oxygen, air, air-water vapor mixture, or enriched oxygen-containing air is introduced to carbon-containing materials such as coal and biomass, which is used to obtain combustible gases such as methane and hydrogen.

Chapter 1 introduces the topic. Chapter 2 proposes a new photocatalyst to replace an organic photocatalyst for hydrogen generation from water and hydrogen sulfide. Chapter 3 discusses various routes to produce solar-based hydrogen, namely, photocatalytic, photo-electrocatalytic, and photobiological decomposition. It also discusses mitigation measures to prevent charge carriers from recombination during the photocatalyst process. Chapter 4 focuses on the use of hydrogen in aviation, the modifications needed to adapt an existing gas turbine to use hydrogen, and a CFD simulation of an existing gas turbine burning hydrogen. Chapter 5 discusses the use of hydrogen oxyfuel in glass and metal industries as a source of heat. The authors suggest an improvement of the exhaust gas stream for energy recovery and to decrease emissions. Moreover, they examine safety issues in the use of hydrogen energy and the acceptance of the community. Chapter 6 studies the effect of hydrogen and carbon monoxide on aluminosilicate refractory. Chapter 7 reviews recent advances in supercritical water gasification of pulping black liquor for hydrogen production. Chapter 8 criticizes minimizing carbon dioxide emissions from coal gasification processes. Finally, Chapter 9 discusses how to improve sustainable hydrogen production yield in hydrothermal gasification processes through novel metal catalysts.

Murat Eyvaz
Associate Professor,
Department of Environmental Engineering,
Gebze Technical University,
Kocaeli, Turkey

Dr. Yongseung Yun
Institute for Advanced Engineering,
South Korea

Dr. Ahmed Albahnasawi
Department of Environmental Engineering,
Gebze Technical University,
Kocaeli, Turkey

Section 1

Hydrogen Energy Processes

Chapter 1

Introductory Chapter: Hydrogen Energy

Ahmed Albahnasawi and Murat Eyvaz

1. Introduction

The purpose of this book is to provide abundant and substantial information about recent development in the production, storage, and application of hydrogen in the energy sector. Since hydrogen energy has benefits in use cases and, unlike synthetic carbon-based fuels, can be truly carbon neutral or even negative throughout its life cycle, hydrogen is emerging as a new energy vector outside of its traditional role and garnering more recognition globally as a viable fuel pathway. This book aims to cover the recent development in the use of hydrogen, storage, transportation, distribution, and the main difficulties and opportunities in the commercial deployment of such systems. It also discusses various methods for producing hydrogen using conventional and renewable energy sources.

2. An overview on energy demand

Because of the ongoing difficulties caused by diminishing fossil fuel supply and deteriorating environmental circumstances, the international community views sustainable development as a long-term issue. Rising energy demands, erratic fossil fuel prices, and significant greenhouse gas (GHG) emissions from fossil fuel-powered vehicles and businesses are the primary causes of this fundamental transition [1–3].

By 2030, the population of the world is expected to reach eight billion, and an increase in energy demand is predicted. In recent decades, renewable energy sources including wind, solar, hydro, and geothermal have drawn a lot of interest. These forms of energy do not produce liquid or gaseous fuels for transportation. Their applicability is constrained by their unpredictable and sporadic existence [4]. Invasive plants and food waste (especially tree trimmings and agricultural crop waste) are additionally low-cost and accessible resources for conversion to clean energy production [5]. Food scraps [6], municipal waste residue, agrochemicals, pharmaceuticals, animal waste, mixed polymers, and lignocellulosic feedstocks are all readily available and inexpensive [7].

Human life, social culture, and economic development all depend on energy. Conventional fossil fuels have been exploited for over two decades, including coal, gasoline, and natural gas, lead to unsustainable oil use, unrestrained exploitation, and major pollution [8]. These non-renewable resources are therefore rapidly approaching degradation and exhaustion [9]. Particularly, the rapid economic transformation and escalating worldwide population increase are driving up energy demand and escalating the energy issue [10]. Furthermore, there is enormous environmental contamination because of the overuse and consumption of fossil fuels. The majority of nations are therefore eager to create a different source of renewable energy [11].

3. Hydrogen energy

Hydrogen is an excellent choice as an energy source for heat and power, among many other uses, due to its many positive qualities, including its overall storage capacity, efficiency, renewability, cleanliness, massive distribution, high conversion, zero emissions, sources, versatility, and quick recovery [12]. It is, therefore, recognized as the most promising and environmentally beneficial energy source of the twenty-first century. It is essential to industrial processes like the generation of ammonia, oil refinement, and water-gas switch reactions [13].

The demand for hydrogen has recently increased dramatically in industrialization's classic oil upgrading industries, including hydro-desulfurization, hydrogenation, and ammonia processing. Due to its affordability and easy access to hydrogen, the majority of the liquid-compressed hydrogen gas used in industrial processing is created commercially using the compression method [14]. Solar power can also be used to produce hydrogen from renewable sources like lignocellulosic biomass or water purification.

3.1 Hydrogen production

Numerous methods exist for obtaining hydrogen from various sources, including electrolysis of water, biofuels, petroleum-based liquids, microbes, and biofuels [14]. Hydrogen production methods and classifications are briefly schematized in **Figure 1**. As seen in the Figure, classification of hydrogen according to the production

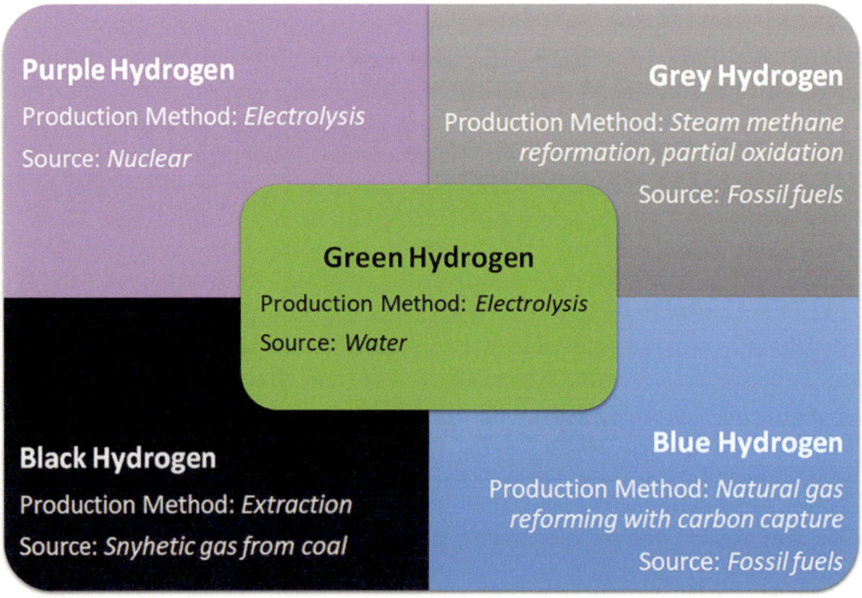

Figure 1.
Hydrogen production methods and hydrogen classification.

method is made with color codes. Greenhouse gas emissions that may occur during the production process are also considered in color coding. For example, hydrogen produced through the electrolysis of water using electricity from renewable sources is coded green and produces near-zero greenhouse gas emissions. Hydrogen, coded in blue, is produced through carbon capture and storage and steam methane reforming. Greenhouse gas emissions from the production of blue hydrogen are characterized as low. Greenhouse gas emissions are high in the gray hydrogen production, produced through steam methane reforming using natural gas [15].

3.2 Hydrogen storage and transportation

There are various methods for storing and transporting hydrogen after its production. The transportation process is done by compressing the hydrogen in gas form or converting it into liquid form in a pressurized environment and then loading it into tankers. However, due to the increasing need for hydrogen in the coming years, it is possible to transport hydrogen through existing natural gas pipelines. For storage, the priority is concentrated on methods that allow transportation. Methods that prioritize transportation for the storage of hydrogen, liquid hydrogen, gaseous hydrogen, metal hydride, and chemical storage. The storage methods of hydrogen are presented in **Figure 2**.

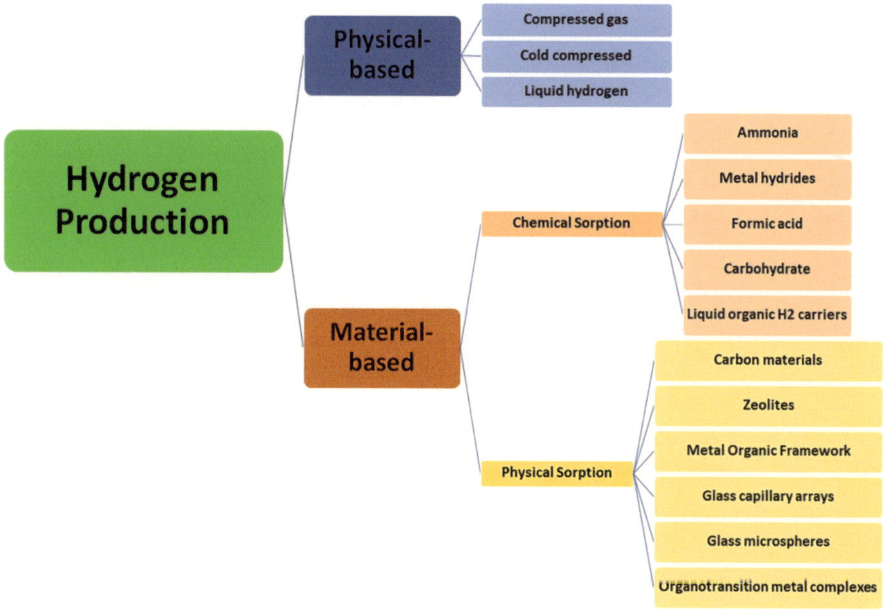

Figure 2.
Hydrogen storage methods (adopted from [16]).

3.3 Hydrogen applications

Hydrogen is an energy storage medium, as well as an energy carrier. It has multiple fuels uses—hydrogen vehicles, stationary power sources, building heating, industrial feedstock, and industrial energy [17] (**Figure 3**). Thus, hydrogen is attracting the

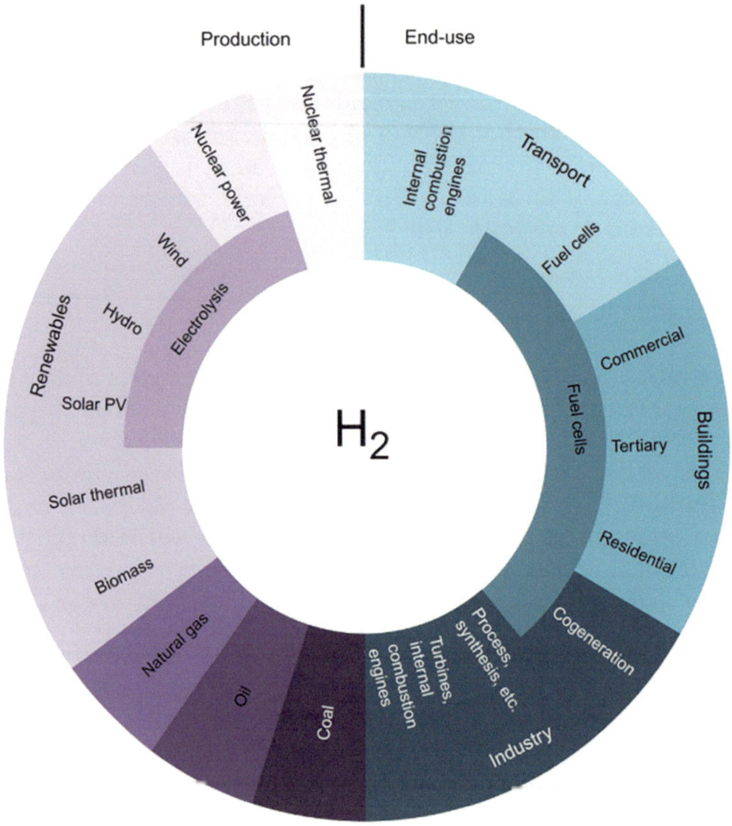

Figure 3.
Production and applications of hydrogen [18].

keen interest of several groups from governments to companies as it offers ways to decarbonize a lot of sectors, especially those that have been proven to be difficult to reduce emissions meaningfully. Moreover, in the industry, hydrogen has had an increasingly important role along the production chain in the last few years [19].

4. Conclusions

Hydrogen energy technologies cannot be used in the industry due to the high production cost, storage difficulties, and transportation costs in today's conditions. However, the fact that hydrogen can be used directly without air pollutants or greenhouse gas emissions and can be produced from low-carbon energy sources are the main reasons for the widespread preference for hydrogen. Clean energy is used to limit air pollution and global warming, especially where the climate crisis has come to the fore in global markets. Various countries and companies see hydrogen as an important clean energy resource that is likely to play a role in the future of the energy sector.

Hydrogen: is a resource that proposes methods that can provide decarbonization in sectors where it is difficult to reduce emissions such as transportation, petrochemistry, and iron and steel. In addition, hydrogen can help improve air quality and increase energy security. Thanks to hydrogen, some technologies can produce, store, transport, and use energy in different routines. Hydrogen, which can be transported in liquid form by pipelines, tankers, and ships, can be converted into electricity and methane, used as energy for households or manufacturing sectors, or as fuel for cars, trucks, ships, and airplanes.

Hydrogen is increasingly forming the basis of mainstream energy debates all over the world. In recent years, many experts have been investigating the potential of using this source to produce, transport, and store hydrogen from various sources and to provide an emission-free final energy supply.

Author details

Ahmed Albahnasawi* and Murat Eyvaz
Department of Environmental Engineering, Gebze Technical University, Kocaeli, Turkey

*Address all correspondence to: albahnasawi@gtu.edu.tr

References

[1] Javaid R, Kawanami H, Chatterjee M, Ishizaka T, Suzuki A, Suzuki TM. Fabrication of microtubular reactors coated with thin catalytic layer (M=Pd, Pd–Cu, Pt, Rh, Au). Catalysis Communications. 2010;**11**:1160-1164. DOI: 10.1016/J.CATCOM.2010.05.018

[2] Rana R, Nanda S, Reddy SN, Dalai AK, Kozinski JA, Gökalp I. Catalytic gasification of light and heavy gas oils in supercritical water. Journal of the Energy Institute. 2020;**93**:2025-2032. DOI: 10.1016/J.JOEI.2020.04.018

[3] Staples MD, Malina R, Barrett SRH. The limits of bioenergy for mitigating global life-cycle greenhouse gas emissions from fossil fuels. Nature Energy. 2017;**22**(2):1-8. DOI: 10.1038/nenergy.2016.202

[4] Nanda S, Azargohar R, Dalai AK, Kozinski JA. An assessment on the sustainability of lignocellulosic biomass for biorefining. Renewable and Sustainable Energy Reviews. 2015;**50**:925-941. DOI: 10.1016/J.RSER.2015.05.058

[5] Okolie JA, Nanda S, Dalai AK, Berruti F, Kozinski JA. A review on subcritical and supercritical water gasification of biogenic, polymeric and petroleum wastes to hydrogen-rich synthesis gas. Renewable and Sustainable Energy Reviews. 2020;**119**:109546. DOI: 10.1016/J.RSER.2019.109546

[6] Yukesh Kannah R, Merrylin J, Poornima Devi T, Kavitha S, Sivashanmugam P, Kumar G, et al. Food waste valorization: Biofuels and value added product recovery. Bioresource Technology Reports. 2020;**11**:100524. DOI: 10.1016/J.BITEB.2020.100524

[7] Nanda S, Berruti F. A technical review of bioenergy and resource recovery from municipal solid waste. Journal of Hazardous Materials. 2021a;**403**:123970. DOI: 10.1016/J.JHAZMAT.2020.123970

[8] Santana KVR, Apolônio FCSO, Wisniewski A. Valorization of cattle manure by thermoconversion process in a rotary kiln reactor to produce environmentally friendly products. Bioenergy Research. 2020;**13**:605-617. DOI: 10.1007/S12155-019-10047-0/TABLES/6

[9] Nanda S, Berruti F. Thermochemical conversion of plastic waste to fuels: A review. Environmental Chemistry Letters. 2021b;**19**:123-148. DOI: 10.1007/S10311-020-01094-7/TABLES/2

[10] Mohr SH, Wang J, Ellem G, Ward J, Giurco D. Projection of world fossil fuels by country. Fuel. 2015;**141**:120-135. DOI: 10.1016/J.FUEL.2014.10.030

[11] Ouyang X, Lin B. Impacts of increasing renewable energy subsidies and phasing out fossil fuel subsidies in China. Renewable and Sustainable Energy Reviews. 2014;**37**:933-942. DOI: 10.1016/J.RSER.2014.05.013

[12] Javaid R. Catalytic hydrogen production, storage and application. Catal. 2021;**11**:836. DOI: 10.3390/CATAL11070836

[13] Sánchez JM, Barreiro MM, Maroño M. Bench-scale study of separation of hydrogen from gasification gases using a palladium-based membrane reactor. Fuel. 2014;**116**:894-903. DOI: 10.1016/J.FUEL.2013.02.051

[14] Singh S, Kumar R, Setiabudi HD, Nanda S, Vo DVN. Advanced synthesis

strategies of mesoporous SBA-15 supported catalysts for catalytic reforming applications: A state-of-the-art review. Applied Catalysis A: General. 2018;**559**:57-74. DOI: 10.1016/J.APCATA.2018.04.015

[15] Scita R, Raimondi PP, Noussan M. Barriers to the Implementation of a Clean Hydrogen Economy; Green Hydrogen: Kolding, Denmark; Milan, MI, USA: Fondazione Eni Enrico Mattei (FEEM); 2020. pp. 8-15

[16] Montenegro AR, Sanjuan M, Carmona M. Energy storage development using hydrogen and its potential application in Colombia. International Journal of Energy Economics and Policy. 2019;**9**:254-268. DOI: 10.32479/ijeep.8294

[17] Sakthivel S. Hydrogen economy offers low-emissions fuel to combat air pollution. Available from: http://gasprocessingnews.com/features/202008/hydrogen-economy-offers-low-emissions-fuel-to-combat-air-pollution.aspx [Accessed: 04 November, 2022]

[18] Mansilla C, Bourasseau C, Cany C, Guinot B, Le Duigou A, Lucchese P. Hydrogen Applications: Overview of the Key Economic Issues and Perspectives. In:Azzaro-Pantel C, editor. Hydrogen Supply Chains (Chapter 7 in Hydrogen Supply Chains: Design, Deployment and Operation). Academic Press-Elsevier; 2018. pp. 271-292. DOI: 10.1016/b978-0-12-811197-0.00007-5

[19] Von Zuben T, Moreira D, Germscheidt R, Yoshimura R, Dorretto D, de Araujo A, et al. Is hydrogen indispensable for a sustainable world? A review of H2 applications and perspectives for the next years. Journal of the Brazilian Chemical Society. 2022;**33**:824-843. DOI: 10.21577/0103-5053.20220026

Chapter 2

Organic Semiconductor for Hydrogen Production

Vikram Rama Uttam Pandit, Shivaji Tulsa Sandu Pandit, Ganesh Punjaba Dawange, Aeed S. Al-Fahdawi, Nurlan Bakranov, Ganesh Kavita Parshuram Jadhav and Omkar Sadhna Arun Malusare

Abstract

The quest of conquering balanced environment for the ultimate search of "Who am I" furnished to pollution and energy crises. As the viable world development is dependent on effective utilization of available renewable energy resources. Hydrogen fuel as an energy source is the future for many upcoming generations as it never produces pollutants. 6, 13 Pentacenequinone (PENQ) is recently developed and reported organic photocatalyst for the generation of hydrogen from water as well as hydrogen sulfide. PENQ can be synthesized and characterized using different methods and techniques/approaches that are listed in this chapter. Green Solid state synthesis method of PENQ is the most promising one as it gives high yield at room temperature and without solvents. Structural characterization of this novel organic catalyst were done using powdered XRD. Cyclic voltammetry is used for the calculating the difference between valance and conduction band levels in the organic PENQ catalyst. After complete structural and morphological characterization, organic PENQ was explored for the hydrogen production from hydrogen sulfide. This photocatalytic nature was also being confirmed using its composites/ coupled systems (PENQ: TiO_2 and PENQ: MoS_2) using hydrogen sulfide and water.

Keywords: hydrogen, organic, photocatalysis, semiconductor

1. Introduction

Considering the utmost importance of Hydrogen (H_2) gas for energy fuel source, scientists has made efforts to find novel ways to get inexpensive H_2 production. Hydrogen generation is the one area of researchers are digging but the storage is the main issue [1]. In this context Titanium dioxide is the very first photocatalyst reported for the production of hydrogen by Gratzel et al. For the improvement in the rate of produced hydrogen there are so many effective improvements are reported like co-catalysts, composites, coupling of two catalyst systems. Metals sulfides and oxides with this type of modifications are also reported [2, 3]. In the category of metal sulfides some binary and ternary sulfides showed excellent result with photocatalytic hydrogen generation. Moreover, graphene based semiconductor photocatalyst materials are also proved best for both generation and storage. As compared to the bulk nanomaterial semiconductor materials are having more surface area, good optical absorption, tunable electronic properties, and lesser charge recombination rate [4–7].

On the other hand, the search of an organic semiconductors for the hydrogen generation is still in progress. Organic five ring system Pentacene (PEN) is well studied semiconductor material in electronic applications. PEN and its derivatives are well utilized in mainly flexible and advanced electronic devices. Recently, it is also showed some very important uses in battery based devices as well as in catalysis [8]. On the other side, PEN and its derivatives are not much stable in air as well as in presence of light [9]. So, unwanted efforts were needed to save it from both light and air. Normally, PEN when come in contact with atmospheric air get oxidize to produce PENQ. This oxidized form of PEN is lately reported as a very important organic semiconductor photocatalyst. As like PEN this oxidized form is also has five aromatic rings with Quinone functional group in the center ring. Interestingly, this PENQ is again reported mostly as a starting compound to get substituted PENQ with fascinating properties [10].

Because of two extra oxygen atoms it gets stabilized and found very effective stability in both air and light. Overall molecular mass, melting point and resistivity towards the acids and other solvents increases. PENQ is with light yellow color and having absorption band edge well in visible region makes it a promising candidate for the photocatalysis applications. Keeping in mind this PENQ was reported for hydrogen generation from hydrogen sulfide gas as well as for the MB dye degradation. After this individual report, it is also found more effective when coupled with inorganic semiconductor materials like TiO2 and ZnO [10–14]. When it combined with other semiconducting materials it produces more hydrogen than the individual one. Yuan et al. very recently proved the effectiveness of this system for the generation of hydrogen from water. These all the modifications on PENQ catalyst is discussed in the present chapter.

By considering the utmost importance of hydrogen energy many attempts were done to produce it using different approaches. Hydrogen energy can be generating from biological, electrolytical, photocatalytical, steam reforming. Out of these photocatalytic method is the simplest and useful to get hydrogen from water and H_2S gas [15–17]. In photocatalysis, photocatalyst is the main hero which alter the rate of hydrogen generation. Plenty of inorganic and organic materials were reported for the production of hydrogen using photocatalysis method.

Also, after the individual organic or inorganic catalysts reports their combinations (composites/hybrid materials) were also reported with enhanced hydrogen production rates. Herein, PENQ and composites synthesis, characterization and photocatalytic activity towards hydrogen production using both water and H_2S were discussed. Enhancement in catalytic activity with the addition of inorganic materials and their effects were also discussed. During the photocatalysis reactions the main role of organic photocatalyst material is to provide necessary charges to complete the reduction reactions. Herein, this main process light is used to stimulate an organic semiconductor material termed as a catalyst which improves the rate of the process [18–20]. Generally, electrons and hole are produced when light absorbed by the semiconductor material.

The photocatalysis mechanism for the organic and inorganic material is the same [21, 22]. In short light is responsible for the traveling electrons from VB to CB [19, 23–26]. Main advantage of organic semiconductors over inorganic semiconductor is that they have high molar absorption coefficient [20, 27].

2. Experimental methods

2.1 Solution synthesis method of PENQ

For the solution method synthesis ortho-phthalaldehyde and cyclohexanedione were mixed thoroughly in an appropriate quantity of ethanol. To this mixture KOH is added dropwise and all the reaction mixture was refluxed till the reddish yellow precipitate get formed. After further purification and drying 40-45% of yield is recorded. In this method elevated temperature is used with organic solvents which is not in accordance to the green chemistry principles and not acceptable.

2.2 Solid-state method synthesis of PENQ

In a typical solid-state synthesis process, phthalaldehyde and cyclohexanedione were ground neatly in a clean mortar and pestle till gel like mass seen. Dropwise addition of KOH is done. During the addition, the brownish solution alters in yellow ppt. After the complete synthesis it was washed with water followed by acetone to remove unwanted products. The formation of PENQ was confirmed by NMR.

2.3 High pressure method synthesis of PENQ

Above aldehyde and ketone were mixed and heated at high pressure using a special autoclave type of reactor. In this method there were no need of any solvent as the reaction products tends to aromaticity and high stability of five membered ring system. As compared to above two methods this present method is found to be more economical and safe.

3. Results and discussion

3.1 PENQ characterization

After complete synthesis of PENQ its structural and morphological characterization were done using these techniques. **Figure 1** depicts the XRD of crystal-line monoclinic PENQ. The XRD diffraction peaks at 9.80, 12.00, 15.01, 23.60 and 27.70° matches with (002), (011), (012), (112) & (104) reflection planes, respectively. For the study of optical properties of PENQ, UV-Visible & Photoluminescence spectra as shown in **Figure 1**. The band gap of PENQ found near 3 eV. Cyclic voltammetry was used for finding the possible band structure of PENQ semiconductor.

A FE-SEM photos in the UV image shows the plate like structure with thickness around 60 nm. There is a stacking of uneven shaped plates and seems to be highly crystalline.

3.2 Photocatalytic hydrogen generation using PENQ

After the possible optical and morphological characterization this PENQ system was utilized for the generation of hydrogen.

Figure 2 depicts the rate of hydrogen production from fresh H_2S gas. Highest produced hydrogen was seen 4848 μmole/h/0.1gm. Same catalyst after proper cleaning and drying were subjected for the same type of experiments showed the almost equal hydrogen generation under the same conditions.

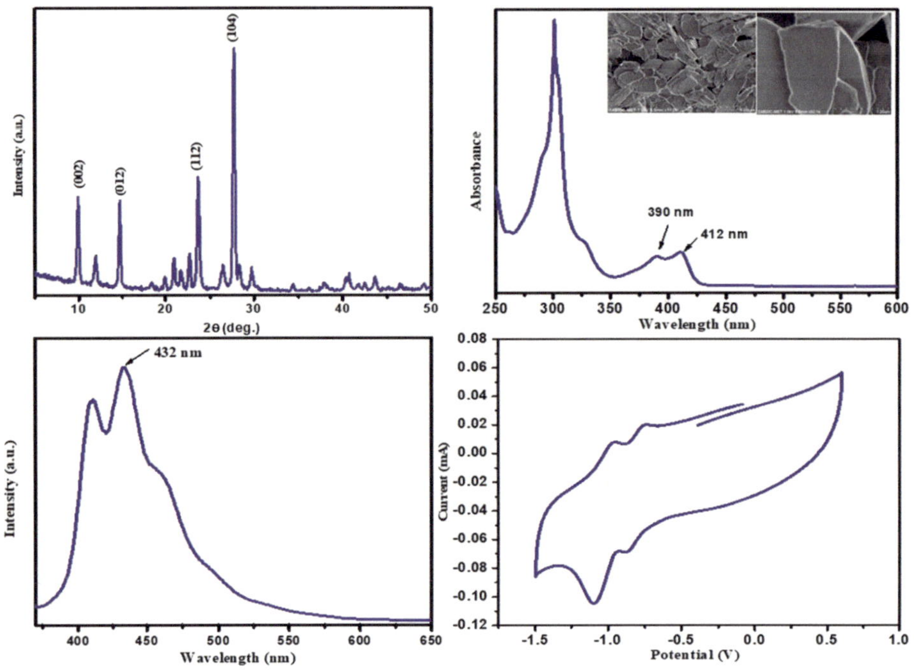

Figure 1.
XRD, UV, PL and CV study of PENQ. Reproduced with copyright permission from American Chemical Society.

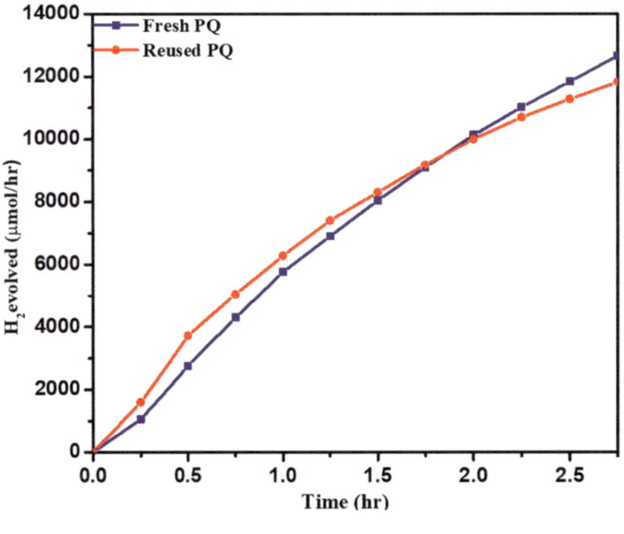

Figure 2.
Photocatalytic Hydrogen generation for PENQ. Reproduced with copyright permission from American Chemical Society.

4. Modification on PENQ with the help of inorganic nanomaterials

After the synthesis and photocatalysis experiments, loan PENQ an organic semiconductor material and by considering photostability and fascinating properties of inorganic TiO_2 an effective and novel organo-inorganic (PENQ: TiO_2) nanosystems is made. All the nanosystems were checked for visible light hydrogen generation experiments. As the stable PENQ has a band gap around 3 eV, it is selected for one of the composite member. In past this PENQ already synthesized using simple methods and it is also having a good molar absorption coefficient. In the present chapter this new organo-inorganic coupled photocatalyst system is prepared using solvothermal reaction method. Along with this characterization of these photosystems were done with the help of UV, PL, SEM, TEM and XRD analysis. Herein, during the individual synthesis of TiO_2 the PENQ is introduced as 5 (PT-5), 10 (PT-10) and 17 mmol (PT-17) of TIP (titanium tetra-isopropoxide) reactant.

4.1 Procedure for the PENQ: TiO_2 nanosystems synthesis

During the solvothermal synthesis method in a clean and dry beaker appropriate amount of titanium tetra-isopropoxide is stirred in methanol and minimum amount of hydrazine hydrate. To the another dry beaker guanidine carbonate is dissolved in acetic acid and added to the first beaker dropwise with continuous stirring for next 30 min. The PENQ which is taken for all the above compositions were taken from earlier reported methods. Further, all the solution is added to the Teflon coated autoclave reactor and kept at 145–150°C around 15-hour reaction time. Lastly, the solids were filtered and dried in vacuum oven. All the photocatalysts systems are well grinded in pestle mortar before the actual photocatalysis experiments.

4.2 Characterization of PENQ: TiO_2 nanosystems

This unique and hybrid nanosystems has been characterized further for phase identification. Powdered X-ray diffractometry (PXRD) tells the story about phase and crystal structure of given materials **Figure 3**. Highlights the PXRD of novel PENQ: TiO_2 photocatalyst nanosystems materials. As the concentration of TiO2 over the surface of PENQ increases the peak intensity is also increases. Both the peaks are highlighted with the help of different shapes in the figure. Diffraction peaks which are present at θ = 25.28, 38.01, 47.9, 54.01 and 54.89 are can be indexed for (101), (004), (200), (105) & (211) planes by corresponding to JCPDS card No. 21-1272 which confirms the Anatase TiO_2. Furthermore, the PXRD peaks θ =14.4, 23.3 & 27.9° are due to the following planes (012), (112) & (104) which matches with the monoclinic crystal structure phase of PENQ. The JCPDS card No.47-2123 is corresponds to monoclinic PENQ in previous reports.

Also, PXRD width in the peaks of inorganic material (TiO_2) highlights the reduction of particle sized to nano level, these sizes are in the range from 5 to 10 nm. After PXRD study same powdered samples were also analyzed using SEM analysis which is shown in the **Figure 2** (a and b). From the SEM pictures it can be clearly seen that these clear PENQ images are having size in the range of 60–250 nm thickness.

Further, SEM pictures of first nanocomposite system for 5 mmol showed the formation very tiny nanoparticles of 5 nm TiO_2 on all over the flakes of organic PENQ sheets. On the other hand, for 10 and 17 mmol nanocomposites systems are showing 6–10 nm sizes, respectively (2 e-h). All the above SEM micrographs shows the uniform stacking of nanoparticles on the surface of sheet like PENQ photocatalyst material. From these images it is observed that the comparative density of nanomaterials is increases as the concentration of TiO_2 is increases.

After the morphological analysis using SEM study nanosystems were subjected to for TEM characterization and SAED pattern analysis as shown in the **Figure 3**. From

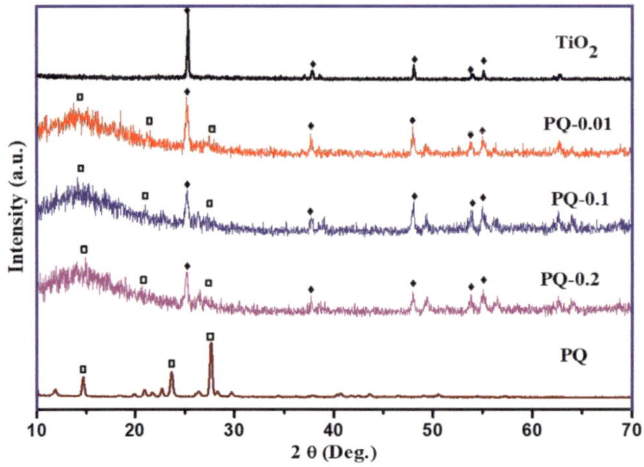

Figure 3.
XRD and UV diffraction analysis. Reproduced with copyright permission from Royal Society of Chemistry.

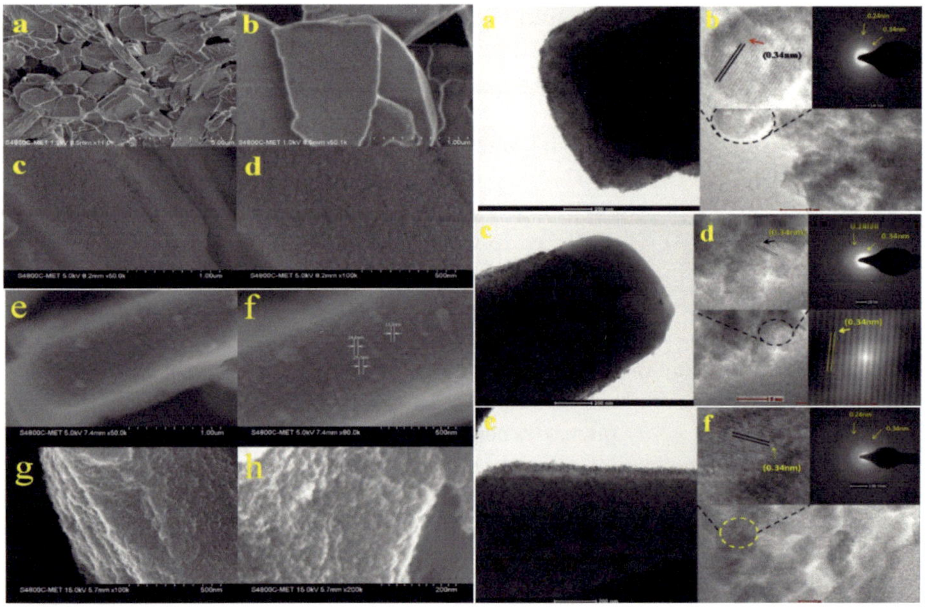

Figure 4.
FESEM and TEM analysis. Reproduced with copyright permission from Royal Society of Chemistry.

the TEM photos it is clear that the nano sized round particles of TiO2 are responsible for the covering of all the sheets of PENQ with 3 nm size. It is observed that at higher concentration of mmol the size of TiO2 is also increases from 3 to 10 nm and covers the sheets like surface of PENQ. Clear and prominent lattice fringes can be seen in high resolution TEM photos (in the **Figure 3b, d** and **f**) with 0.33 nm of inter-planer distance in between the planes corresponds to (101) crystal planes of inorganic nano TiO_2. The ED (Energy Dispersive) patterns clearly shows the high intense rings which also increases from 5 mmol to 17 mmol. When these round particles showed the close connection over the surface gives idea about the effective charge migration mechanism during the catalytic reactions (**Figure 4**).

For optical study diffuse reflectance UV-Visible (UV-DRS) absorbance spectra of PENQ and TiO_2 is given in **Figure 3**. The pure PENQ takes the absorbance edge at 450, 451 attributed the band gap around 2.9 and 2.7 eV. Band gap of TiO_2 in 10 mmol and 17 mmol is high because of blue shift in nano particle.

4.3 Photocatalytic study of PENQ: TiO_2 nanosystem

Semiconductor photocatalysis experiments were performed by manufactured PENQ: TiO_2 systems i.e. PT 5, 10 & 17 mmol from freshly prepared H_2S gas. **Figure 5** shows the band structure and of H_2S gas splitting in presence of UV light. Total hydrogen produced is 36,456 μmol/h/g for PT-5 nanosystem which is higher than PT-10 and 17 catalysts. As compared to TiO2 which is the best photocatalyst, this system produces four times higher hydrogen. 0.25 M KOH at pH 12.5, H_2S splitting follows the as showed mechanism, hydrosulfide HS^- ions (1). This system absorbs the light and produces electron and holes (2). Here holes (h^+_{VB}) are responsible for oxidation HS^- ion in to disulfide ion (S_2^{2-}), taking a proton from HS^- ion (3). Reduction faiths by using electrons to give hydrogen (4).

$$H_2S\,\text{gas} + OH^- \leftrightarrow HS^- + \text{Water}\left(H_2O\right) \tag{1}$$

$$\text{Semiconductor photocatalyst PENQ} : TiO_2 \leftrightarrow h^+VB + e^-CB \tag{2}$$

$$\text{Oxidation process} : 2HS^- + 2\,h^+VB \leftrightarrow S_2^{\,2-} + 2H^+ \tag{3}$$

$$\text{Reduction process} : 2H^+ + 2e^-CB \leftrightarrow H_2 \tag{4}$$

There is no any hydrogen generation was observed without catalyst and without light experiment; this proves that the hydrogen produced is because of PENQ: TiO_2 photocatalyst.

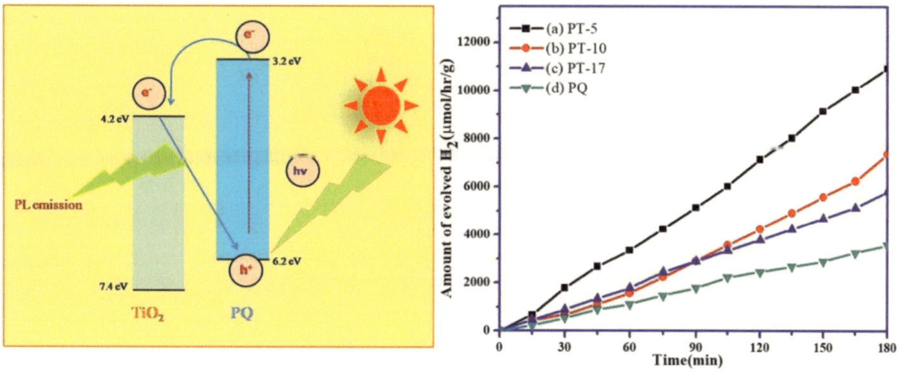

Figure 5.
The band structures and photocatalytic activity of PENQ: TiO_2 hybrid nanosystem. Reproduced by copyright permission from the Royal Society of Chemistry.

5. PENQ: MoS_2 nanosystems for Hydrogen production using water

Lately, PENQ is also used for the production of hydrogen using MoS_2 as a cocatalyst from water. This novel composite system is fully characterized using XRD, Raman spectroscopy, HRTEM, X-ray photoelectron spectroscopy and UV-vis. Here, the role of MoS_2 for the enhancement of PENQ activity was demonstrated effectively towards the water splitting for the faith of hydrogen. Also, the role of MoS_2 (cocatalyst) and TEOA (sacrificial agent) loading effect for the donation of electron was also demonstrated. Precisely, MoS_2 is more effective than Pt as a cocatalyst when combined on PENQ for water splitting.

6. Conclusions

In nutshell, the novel organic PENQ is a very suitable candidate in the race of inorganic semiconductor photocatalysts. PENQ is also showed good agreement in photocatalytical activity towards the degradation of other complex dyes which confirms its activity. The nanosystems is used for the H_2S splitting experiments to get hydrogen using UV-visible light. Highest hydrogen i.e. 36,456 μmol/h/g is seen in absence of cocatalyst. These results give a solid evidence that it can be used with other inorganic nanomaterials to produce hydrogen. It is noteworthy that other hybrid nanosystems with other semiconductor oxide can be synthesized with the present technique. As like PENQ other organic semiconductors may be useful for the construction of novel composite systems. Also, these systems are might have other photocatalytic applications like dye degradation, hydrogen generation and solar cell applications.

Acknowledgements

Authors of the present chapter are thankful to the Principal of H. V. Desai College for continuous encouragement. Dr. V. U. Pandit is also thankful to Savitribai Phule Pune University's (SPPU) Internal Quality Assurance Cell (IQAC) for 'Aspire Research Mentorship Grant' (No. 18TEC001239). Authors are also thanking to Shri. Kiritbhai Shah (Chairman) and Shri. Hemantbhai Maniar (Secretary), 'The Poona Gujarati Kelwani Mandal' (PGKM) for fruitful discussions.

Conflict of interest

Not applicable.

Author details

Vikram Rama Uttam Pandit[1*], Shivaji Tulsa Sandu Pandit[2], Ganesh Punjaba Dawange[3], Aeed S. Al-Fahdawi[4], Nurlan Bakranov[5], Ganesh Kavita Parshuram Jadhav[1] and Omkar Sadhna Arun Malusare[1]

1 Department of Chemistry, The Poona Gujarati Kelvani Mandal's Haribhai V. Desai College, Pune, India

2 Department of Chemistry, Arts, Commerce and Science College, Satral, Ahmednagar, India

3 Sanjivani College of Engineering, Kopargaon, India

4 Department of Chemistry, University of Anbar, Education College for Women, Iraq

5 Research Group AltAir Nanolab, Kazakhstan

*Address all correspondence to: vikramupandit@gmail.com; vikram.pandit@hvdesaicollege.edu.in

References

[1] Kind H, Yan H, Messer B, Law M, Yang PD. Nanowire ultraviolet photodetectors and optical switches. Advanced Materials. 2002;**14**:158

[2] Huang MH, Mao S, Feick H, Yan H, Wu Y, Kind H, et al. Room-temperature ultraviolet nanowire nanolasers. Science. 2001;**92**:1897

[3] Fang X, Bando Y, Liao M, Gautam UK, Zhai C, Dierre B, et al. Single-crystalline ZnS nanobelts as ultraviolet-light sensors. Advanced Materials. 2009;**21**:2034

[4] He ZB, Jie JS, Zhang WJ, Zhang WF, Luo LB, Fan X, et al. Tuning electrical and photoelectrical properties of CdSe nanowires via indium doping. Small. 2009;**5**:345

[5] Chen X, Li C, Gratzel M, Kosteckid R, Mao SS. Nanomaterials for renewable energy production and storage. Chemical Society Reviews. 2012;**41**:7909

[6] Gratzel M. Photoelectrochemical cells. Nature. 2001;**414**:338

[7] Zhai TY, Gu ZJ, Zhong HZ, Dong Y, Ma Y, Fu HB, et al. Design and fabrication of rocketlike tetrapodal CdS nanorods by seed-epitaxial metal-organic chemical vapor deposition. Crystal Growth & Design. 2007;**7**:488

[8] Xie G, Zhang K, Guo B, Liu Q, Fang L, Gong JR. Graphene-based materials for hydrogen generation from light-driven water splitting. Advanced Materials. 2013;**25**:3820

[9] Lee JS, You KH, Park CB. Highly photoactive, low bandgap TiO_2 nanoparticles wrapped by graphene. Advanced Materials. 2012;**24**:1084

[10] Qian W, Greaney PA, Fowler S, Chiu SK, Goforth AM, Jiao J. Low-temperature nitrogen doping in ammonia solution for production of N-doped TiO_2-hybridized graphene as a highly efficient photocatalyst for water treatment. ACS Sustainable Chemistry & Engineering. 2014;**2**:1802

[11] Hirakawa T, Kamat PV. Charge separation and catalytic activity of Ag@ TiO_2 core-shell composite clusters under UV-irradiation. Journal of the American Chemical Society. 2005;**127**:3928

[12] Liu R, Wang P, Wang X, Yu H, Yu J. UV- and visible-light photocatalytic activity of simultaneously deposited and doped Ag/Ag(I)-TiO_2 photocatalyst. Journal of Physical Chemistry C. 2012;**116**:17721

[13] Sawada T, Nakayama S, Kawai-Nakamura A, Sue K, Iwamura H, Hiaki T. Synthesis of polyacenequinones via crossed aldol condensation in pressurized hot water in the absence of added catalysts. Green Chemistry. 2009;**11**:1675-1680

[14] Chun-Lan T, Xu-Hui Z, Mao-Jun D, Yi-Yang L, Shuo S, Gu-Ping O, et al. Synthesis, characterization of the pentacene and fabrication of pentacene field-effect transistors. Journal of the Chinese Physical Society. 2008;**17**(1):281-285

[15] Bhalla V, Gupta A, Kumar M, Rao D, Prasad S. Self-assembled pentacenequinone derivative for trace detection of picric. ACS Applied Materials & Interfaces. 2013;**5**(3):672-679

[16] Bhalla V, Gupta A, Kumar M. Fluorescent nanoaggregates of

pentacenequinone derivative for selective sensing of picric acid in aqueous media. Organic Letters. 2012;**14**(12):3112-3115

[17] Pandit V, Arbuj S, Hawaldar R, Kshirsagar P, Ambekar J, Mulik U, et al. Hierarchical CdS nanostructure by Lawesson's reagent and its enhanced photocatalytic hydrogen production. RSC Advances. 2015;**5**:13715-13721. DOI: 10.1039/C4RA15138K

[18] Pandit V, Arbuj S, Pandit Y, Naik S, Rane S, Mulik U, et al. Solar light driven dye degradation using novel organo–inorganic (6, 13-pentacenequinone/ TiO_2) nanocomposite. RSC Advances. 2015;**5**:10326-10331. DOI: 10.1039/ C4RA11920G

[19] Kumbhar D, Pandit V, Deshmukh S, Ambekar J, Arbuj S, Rane S. Synthesis of hierarchical zno nanostructure and its photocatalytic performance study. Journal of Nanoengineering and Nanomanufacturing. 2015;**5**:227-231. DOI: 10.1166/jnan.2015.1250

[20] Nevase K, Arbuj S, Pandit V, Ambekar J, Rane S. Synthesis, characterization and photocatalytic activity of tungsten oxide nanostructures. Journal of Nanoengineering and Nanomanufacturing. 2015;**5**:221-226. DOI: 10.1166/jnan.2015.1249

[21] Jawale V, Gugale G, Chaskar M, Pandit S, Pawar R, Suryawanshi S, et al. Two-and three-dimensional zinc oxide nanostructures and its photocatalytic dye degradation performance study. Journal of Materials Research. 2021;**36**:1573-1583. DOI: 10.1557/s43578-021-00174-w

[22] Pandit V, Arbuj S, Hawaldar R, Kshirsagar P, Mulik U, Gosavi S, et al. In situ preparation of a novel organo-inorganic 6, 13-pentacenequinone–TiO_2 coupled semiconductor nanosystem: A new visible light active photocatalyst for hydrogen generation. Journal of Materials Chemistry A. 2015;**5**:4338-4344. DOI: 10.1039/C4TA06606E

[23] Pandit V, Arbuj S, Mulik U, Kale B. Novel functionality of organic 6, 13-pentacenequinone as a photocatalyst for hydrogen production under solar light. Environmental Science & Technology. 2014;**48**:4178-4183. DOI: 10.1021/ es405150p

[24] Mazloomi K, Gomes C. Hydrogen as an energy carrier: Prospects and challenges. Renewable and Sustainable Energy Reviews. 2012;**16**:3024-3033. DOI: org/10.1016/j.rser.2012.02.028

[25] Schlapbach L, Züttel A. Hydrogen-storage materials for mobile applications. Nature. 2001;**414**:353-358. DOI: 10.1038/35104634

[26] Andersson J, Gronkvist S. Large-scale storage of hydrogen. International Journal of Hydrogen Energy. 2019;**44**:11901-11919. DOI: org/10.1016/j.ijhydene.2019.03.063

[27] Pandit V. Hydrogen as a clean energy source. IntechOpen. 2021;**1**:1-11. DOI: 10.5772/intechopen.101536

Chapter 3

Production of Hydrogen via Water Splitting Using Photocatalytic and Photoelectrocatalytic Route

Akhoury Sudhir Kumar Sinha, Umaprasana Ojha, Zahoor Alam and Ajay Awdheshprasad Tripathi

Abstract

Hydrogen has been intensively explored recently as an energy carrier to meet the growing demand for green energy across the globe. One of the most difficult and significant subjects in hydrogen energy technology is efficiently creating hydrogen from water by utilizing renewable resources such as solar light. Solar-based hydrogen production comprises several routes, namely, photocatalytic, photoelectrocatalytic, and photobiological decomposition. An efficient photocatalyst is desired to accomplish the above objective by utilizing the first two routes with a minimal rate of recombination of photo-generated charge carriers. In this chapter, strategies for preventing recombination of charge carriers in photocatalysts and the development of photocatalysts have been focused on, and its utilization in the procedure for the production of hydrogen via photocatalytic and photoelectrocatalytic processes is described.

Keywords: hydrogen, photocatalysis, charge recombination, water splitting, photoelectrocatalysis

1. Introduction

Rapid urbanization and global warming have been alarming to people across worldwide in the last decade. This has led to the urgency to develop technology for harnessing green energy from renewable energy resources. Regarding this, researchers have been focusing on the production of hydrogen by utilizing solar energy and abundant water resources. Photocatalytic and photoelectrochemical routes have been widely used for the production of hydrogen by utilizing photocatalysts. Semiconductors-based photocatalysts are essential requirements for the photocatalytic process, and an ideal photocatalyst should have a suitable band positioning and band gap for the utilization of solar energy [1, 2]. Oxide-based and chalcogenide-based semiconductors have been developed for the production of hydrogen in both photocatalytic and photoelectrochemical processes [3, 4]. Swift recombination of photoproduced charge carriers, that is, electron and hole in photocatalyst compromise the durability as well as efficiency of utilization of solar radiations. There are various strategies, such as loading of noble metal, enhancement of crystallinity, morphology variations, coupling semiconductors with others, loading of co-catalysts based on carbonaceous materials, and formation of heterojunction, that have been employed for the prevention of recombination of charge carriers [5, 6].

2. Fundamentals of photocatalysis

Usually, semiconductor materials are used as photocatalysts for the light-induced redox reaction. **Figure 1** illustrates the basic principles of photocatalysis. Semiconductors are characterized by electron structure, namely, valence band (VB) and conduction band (CB). For excitation of charge carrier, when a photon (hv) having energy similar to or higher than the band gap (E_g) of semiconductor falls on the semiconductor, electron migrates from the VB to the conduction band, leaving holes behind. In this excited state, either, electrons of CB and holes in the VB may recombine to dissipate the energy in the form of heat [3].

3. Hydrogen production by photocatalytic water dissociation

The dissociation of water comprises two half-reactions, namely, oxidation and reduction of water to oxygen and hydrogen, respectively. The following equations refer to the dissociation of water into oxygen and hydrogen in alkaline media:

Cathode

$$2H_2O + 2e^- \rightarrow H_2 + 2OH^- \quad (1)$$

Anode

$$2OH^- \rightarrow \frac{1}{2}O_2 + H_2O + 2e^- \quad (2)$$

Overall

$$H_2O \rightarrow \frac{1}{2}O_2 + H_2 \quad \Delta G = 238\ kJ/mol \quad (3)$$

$$\Delta G = -nFE \quad (4)$$

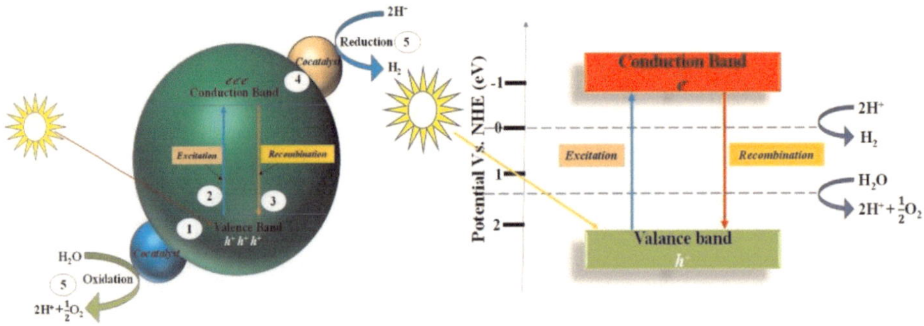

Figure 1.
The basic principle of semiconductor-based photocatalysis ref. [3].

where n stands for the number of electrons associated with the photocatalytic reaction, F is the faraday constant, and E is the redox potential energy of the reaction.

Thermodynamically, the change of Gibbs free energy is positive according to eq. (3); therefore, energy is needed for the dissociation of water. In other words, dissociation of water is not a spontaneous process. The minimum energy required per electron will be 1.13 eV from eq. (4), which is equal to the photon energy of wavelength 1100 nm according to the eq. (5)

$$E = 1240/\lambda \text{ eV} \tag{5}$$

A photon that has an energy of 1.23 eV or greater shall be able to dissociate water. Solar radiation may be utilized as it has a high quantum of photons below the wavelength of 1100 nm. It is important to note that light does not dissociate the water because water is transparent to light. Therefore, a medium (semiconductor catalyst) is required to induce water splitting. A semiconductor has a narrow band gap, where electrons are generated by the utilization of photon of wavelength less than 1100 nm. Moreover, the conduction band gap of catalysts may be preferably more negative than the reduction potential of H_2O to initiate the water-splitting reaction by an electron. Holes at the valence band are used by the sacrificial agents present in the electrolytes.

The overall photocatalytic dissociation of water consists of mainly three steps (as shown in **Figure 2**): (1) electron–hole pair's generation by using the absorption of a photon, (2) electron–hole separation and migration of electrons to the conduction band, (3) reactions between surface species and electron. The first step is photon-physical processes, the second step is the migration of charge carriers, and the third

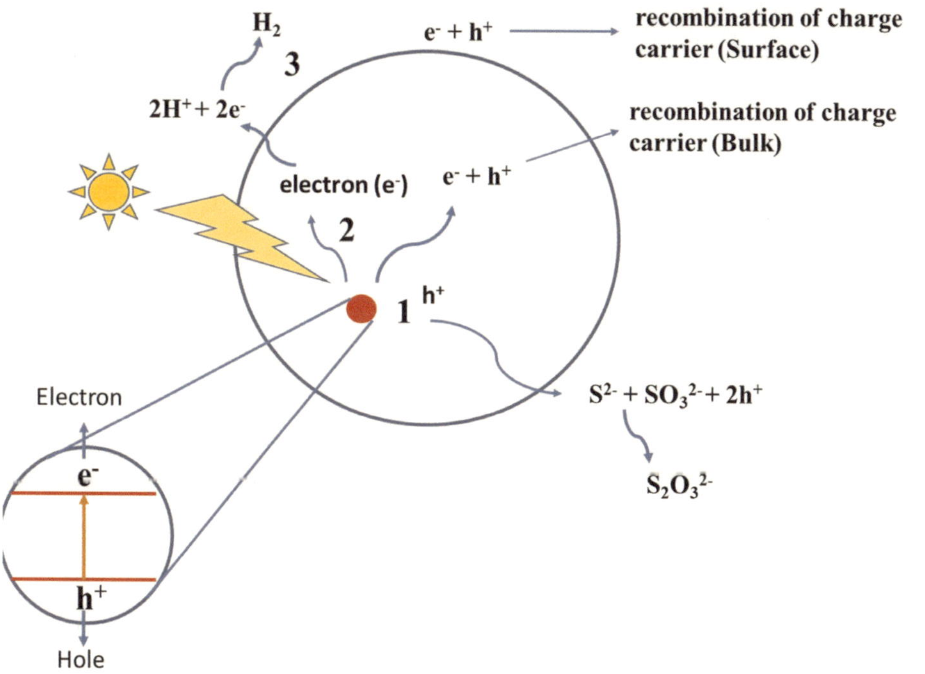

Figure 2.
Fundamental steps associated with the photocatalytic water dissociation of water for hydrogen.

step is the chemical process. Moreover, surface and bulk recombination of electron–hole may also occur. The reverse reaction is also possible in water. Therefore, research is undertaken toward (a) improvement in the harvesting of visible light by utilizing semiconducting materials and (b) efficient charge separation tools and suppressing electron and hole recombination. Apart from these, it is also required to make modifications to the particle morphology of the photocatalyst and adopt the surface structure modification to enhance the rate of the reaction of redox couples.

4. Photoproduced charge recombination phenomenon in photocatalyst

Fast recombination of photoproduced electrons and holes diminishes photocatalytic efficiency, as is well known. The recombination of the photoproduced electron and holes happens in the nanosecond (1×10^{-9} second) range due to a high Coulomb constant (8.99×109 N m^2 C^{2-}) [7]. As a result, preventing the recombination of photoproduced electrons and holes is a more difficult and daunting task. Various tactics for preventing the recombination of photo-induced electron and hole pairs have been proposed in this area, including the loading of noble metals and the coupling of two semiconductors resulting in the creation of a heterojunction at the interface. In general, a heterojunction is a contact between two semiconductors with differing structures. It effectively isolates photo-induced electrons and holes in photocatalysts, increasing photocatalytic efficiency overall.

4.1 Doping of noble metals on the semiconductor surface

For the avoidance of photo-induced electron and hole recombination in semiconductors, loading metal or noble metal on the surface of the semiconductors as co-catalysts has been extensively researched. A Schottky barrier is generated at the interface between the active metal and semiconductors because noble metal has a larger work function than semiconductors. In the literature, the creation of the Schottky barrier has been extensively researched [8]. When noble metals with higher work functions (m) than semiconductors with lower work functions (s) are electrically connected (**Figure 3**), electrons migrate from the conduction band of the semiconductor surface to the active metal until their Fermi levels are aligned, forming a

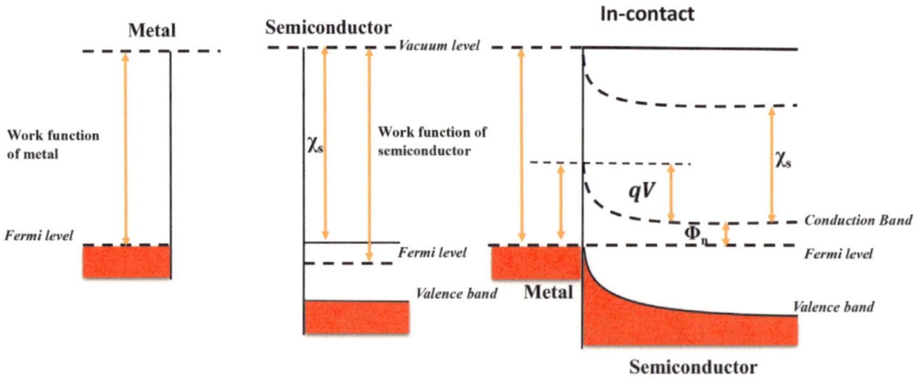

Figure 3.
Energy level diagram of metal surface with n-type semiconductor possessing $\phi_m > \phi_s$ before and after the junction (ref. [9]).

space charge region at the interface between them. As a result, the noble metal will accumulate an excessive amount of negative charge on its surface, whereas the semiconductor will accumulate a positive charge as a result of electron transport. Furthermore, this causes an upward band bending of semiconductors toward the surface. As a result, the Schottky barrier is a tiny barrier that forms in the depleted region.

4.2 Coupling of semiconductors with one another

Making a composite of two or more semiconductors implanted with additional semiconductors with appropriate band energies is another way to boost its photocatalytic activity and stability. According to the literature, a heterojunction formed by the conjunction of two semiconductors can be used to separate charges. Owing to the diffusion of electrons and holes, a p–n junction with a space-charge zone may occur when the p- and n-type semiconductors come into contact (**Figure 4**). As a result, a built-in electrical potential is established, which allows electrons and holes to flow in the opposite direction. Due to the diffusion of electrons and holes, a p–n junction with a space-charge zone may emerge at the interfaces. As a result, a built-in electrical potential is produced, allowing electrons and holes to flow in the opposite direction. The photoproduced electron–hole pairs can be promptly separated by the induced electric field within the space charge area when the p–n heterojunction is irradiated by photons with energy greater than or equal to the photocatalyst's band gap. This p–n heterojunction has several advantages: (a) efficient separation of charge; (b) swift charge migration to the active surface; and (c) sustainable charge carrier life. All of these characteristics impact the photocatalytic performance of the p–n heterojunction [10].

4.3 Creation of heterojunction

As we know that recombination of photoproduced charge carriers in a photocatalytic water-splitting process can be considered a devil to hinder the production of hydrogen at a large scale by utilizing ample solar energy. Prevention of charge recombination is a bit challenging and overwhelming task. Regarding this, the creation of heterojunction between the semiconductors at the interfaces is considered a

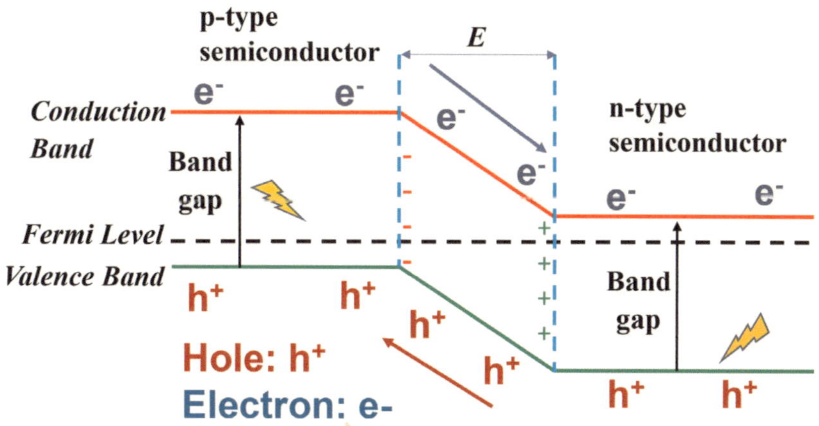

Figure 4.
Schematic representation displaying the band structure and charge carrier separation in the p–n heterojunction ref. [10].

prominent strategy for the designing of efficient photo-catalysts, which could enhance the charge separation efficiency. Low et al. have reported five types of heterojunction in their review literature, namely, convention type-II heterojunction, p-n type heterojunction, surface heterojunction direct Z-scheme heterojunction, and recently developed semiconductor-graphene heterojunctions (SC-Graphene) [9].

5. Advancements in photocatalysts in water splitting

The significant optoelectrical properties, such as electrical conductivity, band gap, and absorption of light, are required for particulate semiconductors-based photocatalytic water splitting. Over four decades, various particulate-based-semiconductor photocatalysts have been developed, namely, oxide-based photocatalysts (TiO_2, ZnO, $BiVO_4$, $SrTiO_3$, etc.) and chalcogenide-based-photocatalysts (CdS, ZnS, MoS_2, CdSe), and are widely used for hydrogen generation based on the photocatalytic water splitting utilizing visible light. However, it has been reported that the fast recombination of photo-induced electrons and holes limits the production of hydrogen at a higher scale. Shao et al. [11] have developed carbonized MoS_2/Mo_2C-decorated CdS with an extraordinary quantum efficiency (41.4% in presence of 420 nm) and reported 112 times greater activity compared to that of the CdS. The exceptional photocatalytic performance of the above system is assigned to the presence of the metallic character of MoS_2 and Mo_2C and an appropriate Gibbs free energy (ΔG) of H_2 adsorption, resulting in increased absorption of photon, rapid separation, and migration of photo-induced carriers, and displaying optimum activity in the HER. Guncai et al. [12] employed the dual-band-gap strategy and prepared CdS-supported reduced graphene oxide (RGO) nano-heterostructure and investigated that interfacial interaction between CdS facilitates the fast transfer of electrons from the CdS to RGO. This interfacial interaction, dependent on the transfer of charge carriers, has been proved theoretically as well as experimentally. Moreover, the transfer of single electron and holes present on the surface of catalysts depends on the surface potential. Surface potential enhances the transfer of electrons to the surface of RGO. Guoning et al. [13] emphasized the stability of heterojunctions between CdS-Cu_{2-x}S and MoS_2 by the migration of ion, that is, Cu, and reported that intercalation of Cu within the MoS_2 basal plane enhances the epitaxial growth of CuI@MoS_2 nanosheets in a vertical manner on the surface of 1D core–shell CdS-Cu_{2x}S type nanorods to generate catalytic and protective layers, enhancing catalytic activity and durability at the same time. This design concept shows how hybridized surface layers can be used as efficient catalytic and shielding interfaces for photocatalytic hydrogen generation. Jing et al. [14] fabricated the few-layered MoS_2/ZnCdS/ZnS at higher temperatures and constructed the dual heterostructures, which showed a higher rate of production of hydrogen as a result of preventing the recombination of photoproduced electrons and holes. The well-designed structure is effective in the separation and migration of photoproduced electron–hole pairs, leading to an enhancement in photocatalytic H_2 production activity, according to time-resolved photoluminescence spectra. Menglong et al. [15] prepared carbon nitride (CN)-supported red phosphorous (RP) and reported that RP tightly bound with the carbon nitride (CN) framework. A van der Waals heterojunction has been formed between CN and RP and promoted the rapid migration of electrons for the conversion of H^+ ions to H_2 in the process of solar-driven water splitting. Van der

Catalyst	Preparation Method	Light Source	H_2	Ref.
$TiO_2/MoS_2/$ graphene	Hydrothermal MoS_2/graphene Composite prepared. Composite and $Ti(OC_4H_9)$ are reacted hydrothermally to obtain a catalyst	350 W Xe arc lamp	165.3 μmol h^{-1}	[18]
RuDTC: Pt@CdS	CdS nanorods were prepared by seed growth method on it Pt nanoparticle growth was done. Composite of Pt@CdS on its, Ru complex has grown as chemical ligation	$20mWcm^{-2}$	16 μmol h^{-1}	[19]
CdS Quantum Dots/ZnO	CdS QDs on ZnO (microflowers) by in situ deposition using the SILAR method	225 W xenon arc lamp	22.12 mmol/h*g	[20]
$NiS_{1.97}$	NiO reacts with Thioacetamide in presence of Inter atmosphere (Ar gas)	solar simulator 100 mW/cm^2	~4 μmol h^{-1}	[21]
$La_2Ti_2O_7/CuO$	$La_2Ti_2O_7$ and $Cu(NO_3)_2.3H_2O$ in a crucible for the calcination process.	125 W medium pressure Hg lamp	936.6 mmol h^{-1}	[22]
Se Incorporation in $La_5Ti_2CuS_5O_7$	$La_5Ti_2CuS_5O_7$ and $La_5Ti_2CuS_5O_7$:Se both were prepared by the reaction in a solid state	300 W Xe lamp	~175 μmol h¯1	[23]
Palladium Sulfide (PdS)	Pd films in vacuum-sealed ampules with sulfur powder for heating	200 W halogen lamp	~0.55 μmol min^{-1}	[24]
$ZnIn_2S_4$	$Zn(CH_3COO)_2{\cdot}2H_2O$ and thioacetamide hydrothermally give a product	FX-300 Xe lamp	100.1 μmol h^{-1}	[25]
$CdS\text{-}Cu_{2-x}S$	Solvothermal or colloidal method CdS and catalyst preparation by a cation exchange process	300 W xenon lamp along with U.V. cut-off filter	14184.8 μmol $g^{-1} h^{-1}$ (20 h)	[13]
$MoS_2/g\ C_3N_4/$ GO	Graphene oxide is prepared by exfoliation MoS_2 by the reported method and $g\text{-}C_3N_4$ too. Composite preparation by anion exchange and hydrothermal.	300 W xenon arc lamp	1.65 μmol h^{-1}	[26]
$CdS/WO_3/$ $CdWO_4$	CdS prepared first and then WO_3 and $CdWO_4$ three of them are reacting to make tricomposite.	Xe lamp	~3100 μmol h^{-1}	[27]
$La_{0.5}\ Sr_{0.5}\ Ta_{0.5}$ $Ti_{0.5}O_2N$	Synthesized by nitridation of oxide resulted from a polymerizable complex method	Xe arc lamp (300 W)	18.2 μmol	[28]
$(CuGa)_{0.5}\ ZnS_2$	$(CuGa)_{0.5}ZnS_2$ synthesis by Vacuum solid-state reaction method	Xe arc lamp (300 W)	25.6 μmol h^{-1}	[29]
$CuGaS_2$/RGO-TiO_2	$CuGaS_2$ composite on RGO-TiO_2	Xe arc lamp (300 W)	6.9 μmol h^{-1}	[30]
Pt/$CuGaS_2$ and RGO-TiO_2	Pt-loaded $CuGaS_2$ composite on RGO-TiO_2	Xe arc lamp (300 W)	19.8 μmol h^{-1}	[30]
Zn_xCd_{1-x} Se Coated with $[Fe(CN)6]^{4-}$	Preparation of the material is via the solid-state method in a quartz ampoule which is sealed	Xe arc lamp (300 W)	0.75 μmol h^{-1}	[31]
CdSe-DF and CdSe-NR (*dendritic fractals and nanorods)	CdSe is prepared by liquid-solid growth mechanism in hydrothermal for the getting different morphologies, on it, Cu_3P is incorporated.	500 W Xe-Hg lamps	46.5 mmol/h/g (CdSe-DF)	[32]

Catalyst	Preparation Method	Light Source	H_2	Ref.
CdSe-QD/amp-TiO_2	CdO, oleic acid, and trioctylamine were taken in three-necked flasks, degassed under vacuum, simultaneously that Se and trioctylphosphine (TCP) mixture were prepared inside Ar filled glove box. Quickly inject TCP-Se into a three-necked flask at 3000C	AM 1.5 light irradiation	436 μmol/g.h	[33]
$ZnIn_2S_4/MoSe_2$	$ZnIn_2S_4$ prepared in Autoclave. $ZnIn_2S_4/MoSe_2$ composites prepared by solution-phase hybridization method.	300 W high-pressure Xe arc lamp	2228 μmol/g.h	[34]
MoS_2 -QDs/g-C_3N_4	MoS_2 quantum dots were prepared by bath and probe sonication. MoS_2-QDs/g-C_3N_4 composite prepared by sonication and after centrifugation	300 W Xe lamp	19.66 μmol h^{-1}	[35]

Table 1.
Photocatalytic activities of various photocatalysts.

walls heterojunction is one of the best types of heterojunction that could increase the efficiency of the photocatalytic H_2O splitting route. Kun et al. [16] synthesized CdS supported on the nanosize MoS_2–graphene hybrid. They reported that this catalyst showed a significantly higher hydrogen production. They also optimized the quantity of nanosize of MoS_2–graphene hybrid and found that the highest photoactivity was observed when 2 wt% of MoS_2–graphene was used having MoS_2: graphene = 1:2 (mol:mol). The remarkable activity of the prepared photocatalyst was ascribed to the higher rate of absorption of H^+ on the unsaturated S atom at the edge of MoS_2. It is believed that the presence of graphene could separate the photoproduced electrons and holes and thus retard the recombination of charge carriers. Xiaohu and coworkers [17] also synthesized rGO–CdS by utilizing the hydrothermal procedure using Cd precursor, GO, and $Cd(Ac)_2.2H_2O$ as a precursor in DMSO. It was sonicated and then this suspension was poured into a stainless-steel autoclave with a Teflon lining for further solvothermal treatment at 453 K. They also prepared rGO–CdS by a precipitation process for comparison. The composite exhibited improved photocatalytic activity toward H_2 production compared to that of the CdS, whereas the physical mixtures of rGO and CdS failed to display a similar enhancement in efficacy. The improved activity was assigned due to the incorporation of rGO that promoted suppression of recombination of the photoproduced carrier as an electron-transfer channel and acceptor, as well as a support of CdS in an aqueous solution.

It is well known that heterojunctions, dual-band structure, van der Walls heterojunction, dual heterojunction, and precipitation method could directly influence the activity. These strategies are the advancement in photocatalytic water splitting. The activities of various photocatalysts, preparation methods, and sources of light are reported in **Table 1**.

6. Factors affecting photocatalytic water splitting

Apart from the fabrications and prevention of recombination of photoproduced electrons and holes, various significant factors are also affecting the photoactivity of photocatalysts such as band gap energy, light intensity, phase of the catalyst, sacrificial agents, and advancement of photoreactor design.

6.1 Band gap

The electronic configuration of semiconductors-based photocatalysts are generally described in term of energy band and can be represented by the difference in energy level by conduction band (CB) and valence band (VB). In general, the acceptor's relative potential level must be less than the semiconductor's CB for thermodynamic reasons. Moreover, a more negative value of the conduction band position is favorable for the reduction of H^+ ions to hydrogen gas (H_2) [36].

6.2 Intensity of light

Concentration of light possessing energy higher than the activation threshold can boost photocatalytic water-splitting efficiency. In terms of UV photon flux, there are two systems for photocatalytic splitting. Chemical processes consume electron–hole pairs quicker than recombination reactions in the first-order regime, which is typical for fluxes from small-scale experiments of 25 $mWcm^2$. The second-order regime is for greater intensities, where the rate of recombination is normally dominating and has a smaller effect on the reaction rate. Furthermore, the variation of reaction rate in the water-splitting process is considered to be dependent on the wavelength that comes after the photon absorption of the catalyst with a threshold value corresponding to the band gap of the photocatalyst (**Figure 5**) [36].

6.3 Effect of temperature

Thermodynamically, the temperature did not induce the photocatalytic activity because it cannot generate the electron and holes in catalysts. However, the effect of temperature can be generally seen as facilitating the release of the H_2 from the surface

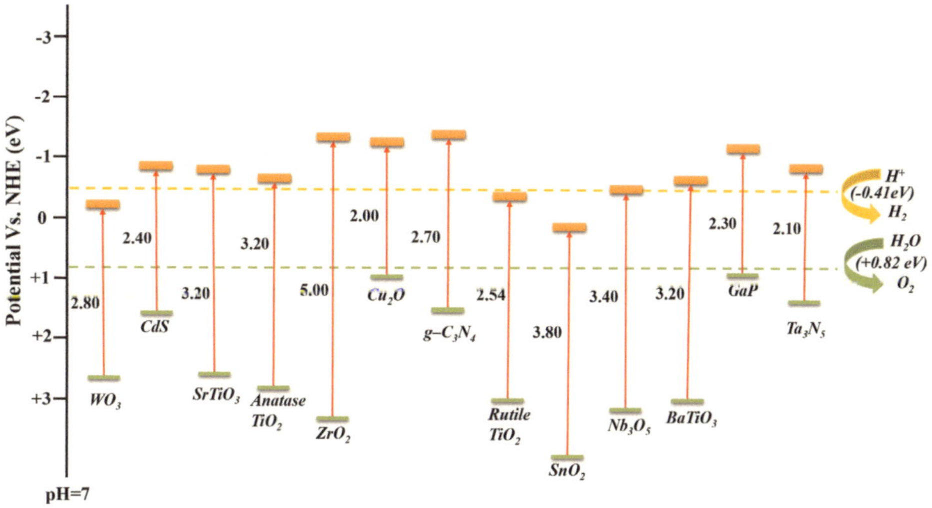

Figure 5.
Band structure of different types of photocatalysts (ref. [36]).

of photocatalysts, thus increasing the activity of photocatalytic water splitting. Moreover, the temperature increases the rate of reactions.

6.4 pH

It is well known that the rate of H_2 production from H_2O splitting is dependent on the pH of the solution, that is, hydrogen ions concentration. The hydrogen production is to be generated more effectively in a weak base solution rather than the stronger base solution (pH > 10) or in acidic. Sinha et al. have optimized the pH of the solution (8.6) for a high rate of hydrogen production in the sacrificial agent-mediated photocatalytic reactions [37].

7. Photoelectrochemical water splitting

Photoelectrocatalyst is a material that has the capability of harvesting solar light and chemical transformation by an electrochemical redox reaction. The first photoelectrochemical route for studying water splitting was performed by Fujishima and Honda using TiO_2 catalyst [38, 39]. Photoelectrocatalytic water-splitting reaction is performed in a photoelectrochemical cell. The photoactive material is the semiconductor material that is used for water splitting. In a photoelectrochemical cell setup, which is very similar to an electrochemical cell, either one or both electrodes is photoactive material that is coated on a thin-film-conducting material like ITO/FTO, etc. (**Figure 6**). The photoelectrochemical way of water splitting has an advantage over photocatalytic water splitting, such as the facility of gaseous product separation and external bias voltage that can operate reaction kinetics and mechanism for product choosy. The appealing superiority of photoelectrochemical water splitting takes place at a semiconductor surface, which can generate an electric field at a semiconductor-liquid junction [40]. For the last three decades, the photoelectrochemical route is being used for the generation of hydrogen. Activities of different photoelectrocatalysts are reported in **Table 2**.

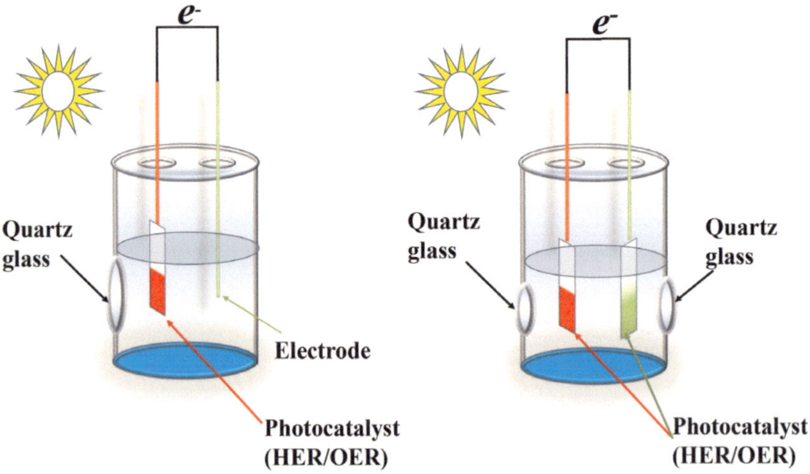

Figure 6.
Schematics of photoelectrochemical cells.

DOI: http://dx.doi.org/10.5772/intechopen.106915

Catalyst	Preparation Method	Light Source	H_2	Ref.
n-$La_2Ti_2O_7$/ p-$LaCrO_3$	$La_2Ti_2O_7$ is prepared by the calcination process and $LaCrO_3$ is also prepared by the same method as above	125 W medium pressure Hg lamp	267.6 μmol h^{-1}	[41]
Pt-Au/g-C_3N_4	g-C_3N_4 is prepared by heating melamine at 5500°C. Pt-Au/gC_3N_4 is prepared in a Teflon-lined stainless-steel autoclave by putting a solution of Pt and Au salt and gC_3N_4.	300 W Xe lamp	267.6 μmol .g^{-1}.h^{-1}	[42]
$La_2Ti_2O_7$/CuO	$La_2Ti_2O_7$ and $Cu(NO_3)_2.3H_2O$ in a crucible for the calcination process.	125 W medium pressure Hg lamp	936.6 mmol (in 4 h)	[22]
Palladium Sulfide (PdS)	Pd films in vacuum-sealed ampules with sulfur powder for heating	200 W halogen lamp	~0.55 μmol. Min^{-1}	[24]
$ZnIn_2S_4$	$Zn(CH_3COO)_2{\cdot}2H_2O$ & thioacetamide hydrothermally give a product	FX-300 Xe lamp	100.1 mmol. h^{-1}	[43]
CdS-Cu_{2-x} S	Solvothermal or colloidal method CdS and catalyst preparation by a partial cation exchange reaction	300 W xenon (U.V. cut-off filter)	14184.8 μmol. $g^{-1}{\cdot}h^{-1}$ (20 h)	[13]
MoS_2/g C_3N_4/GO	Graphene oxide is prepared by exfoliation MoS_2 by the reported method and g-C_3N_4 too. Composite preparation by anion exchange and hydrothermal.	300 W xenon lamp	1.65 mmol. h^{-1}	[44]
$(CuGa)_{0.5}ZnS_2$	$(CuGa)_{0.5}ZnS_2$ synthesis by Vacuum solid-state reaction method	300 W Xe lamp	25.6 μmol. h^{-1}	[29]
$Zn_xCd_{1-x}Se$ Coated with $[Fe(CN)_6]^{4-}$	Preparation of the material is via the solid-state method in a quartz ampoule which is sealed	300 W Xe lamp	0.75 μmol. h^{-1}	[45]
CdSe-QD/ amp-TiO2	CdO, oleic acid, and trioctylamine were taken in three-necked flasks, degassed under vacuum, simultaneously that Se and trioctylphosphine (TCP) mixture were prepared inside Ar filled glove box. Quickly inject TCP-Se into a three-necked flask at 3000°C	AM 1.5 light irradiation	436 μmol/g.h	[46]
$ZnIn_2S_4/MoSe_2$	$ZnIn_2S_4$ preparation in Autoclave which Teflon stainless steel. $ZnIn_2S_4/MoSe_2$ composites preparation by solution-phase hybridization method.	300 W high-pressure Xe arc lamp	2228 μmol/g^{-1}.h	[34]
CoSx@POP (POP is a porous organic polymer)	7CoSx@POP preparation involves three steps first is POP preparation by oxidative polymerization method second is Co_3O_4@POP preparation which by heating POP and $Co(OAc)_2$ for 12 hr. in an oil bath and the third step is catalyst preparation	300 W Xenon lamp (420 nm cut-off)	180 μmol/cm^2 after 120 min	[47]
Cu_2S-TiO_2/TiO_2 - NP/FTO	Firstly, a layer of TiO2 grown on FTO glass and, then Cu_2S-TiO_2 grows on TIO_2/FTO	100 W xenon arc-lamp	~ 55 ± 5 μmol/ cm^2hour	[48]
Cu Doped ZnS/ZnO	CuO/ZnO Prepared by solvothermal and catalyst prepared by hydrothermal method	300 W Xe lamp (cut-off filter)	1.044 mmol g^{-1} h^{-1}	[49]

Catalyst	Preparation Method	Light Source	H_2	Ref.
g-C_3N_4/$BiVO_4$	Prepared by hydrothermal for $BiVO_4$ and thermal polycondensation of urea for g-C_3N_4	500 W halogen lamp	21.4 mmol/h	[50]
Mo-In_2O_3–$ZnIn_2Se_4$	The catalyst prepared by $Zn(NO_3)_2 \cdot 6H_2O$, $In(NO_3)_3 \cdot xH_2O$, and Na_2SeO_3 by solvothermal reaction	300 W Xe lamp	~1200 μmol/g	[51]

Table 2.
Photoelectrocatalytic activities of various catalysts.

Overall, the following inference can be drawn from the results described in the above section.

- Possibly, chalcogenide based-photocatalysts give higher activities under irradiation of visible light due to the presence of conduction band position toward more negative potential with respect to NHE.
- Reported results show that higher production of hydrogen is seen due to the formation of heterojunction of multiple active catalytic systems.
- Oxide-based photocatalysts have limitation to maximize hydrogen production due to the recombination of photoproduced charge carrier. This may be facilitated by UV light, which comprises ~4–5% of the solar spectrum.
- Results also revealed that the presence of two-dimensional (2D) co-catalysts, such graphene, MoS_2, and C_3N_4 with semiconductors photocatalysts, results in higher activities.
- Nanocatalysts with precise nonmetallic doping may also be helpful to drive the efficiency in a positive direction.

8. Conclusions

Photocatalytic and photoelectrocatalytic routes are considered as sustainable technology for the generation of green and clean hydrogen from renewable sources. Advancements have been made in photocatalysts to prevent the recombination of photoproduced charge carriers, that is, electron and hole. But still both the routes are operational at Research & Development level, and a ground-breaking research is needed in terms of the fabrication of durable and sustainable photocatalysts, development of photoreactor. The catalyst heterostructure, size, shape, and composition controlled the H_2 production efficiency under photocatalytic conditions. The availability of photocatalysts with higher efficiency and durability is also desirable for the said purpose. Subsequently, the dispersion of the catalysts in the aqueous medium and their exposure to light during the photocatalysis process is important for large-scale operations. The ability of photocatalysts to produce H_2 from low-grade water, wastewater, and sea water is envisaged to further enhance the commercial viability of the process. Other avenues such as photoelectrocatalysis may also be explored to understand the commercial viability of these routes.

Acknowledgements

The authors are thankful to the editor and reviewers for their valuable comments and observations.

Conflict of interest

The authors declare no conflict of interest.

Nomenclature

$h\nu$	Energy of photon
VB	Valence band
CB	Conduction band
E_g	Band gap
ΔG	Gibbs free energy
n	Number of electrons
F	Farady constant
λ	Wavelength
N	Newton
m	meter
C	Coulomb
Φ_m	Metalwork function
Φ_s	Semiconductor work function
E_f	Fermi level
ϕ	Work function
s	Electron affinity

Author details

Akhoury Sudhir Kumar Sinha*, Umaprasana Ojha, Zahoor Alam
and Ajay Awdheshprasad Tripathi
Rajiv Gandhi Institute of Petroleum Technology, Jais, Amethi, India

*Address all correspondence to: asksinha@rgipt.ac.in

References

[1] Sugang M, Jinfeng Z, Shifu C, Sujuan Z, Weixin H. Perspective on construction of heterojunction photocatalysts and the complete utilization of photoproduced charge carriers. Applied Surface Science. 2019; **476**:982-992. DOI: 10.1016/j.apsusc.2019.01.246

[2] Hui P. Principles on design and fabrication of nanomaterials as photocatalysts for water-splitting. Renewable and Sustainable Energy Reviews. 2016;**57**:584-601. DOI: 10.1016/j.rser.2015.12.117

[3] Xiaobo C, Shaohua S, Liejin G, Samuel SM. Semiconductor-based photocatalytic hydrogen generation. Chemical Reviews. 2010;**110**(11): 6503-6570. DOI: 10.1021/cr1001645

[4] Sha C, Danlian H, Piao X, Wenjing X, Lei L, Min C, et al. Semiconductor-based photocatalysts for photocatalytic and photoelectrochemical water splitting: Will we stop with photocorrosion? Journal of Materials Chemistry A. 2020; **8**:2286-2322. DOI: 10.1039/C9TA12799B

[5] Weiwei H, Lulu C, Weiyu S, Shaobin W, Xiaobin F, Li Y, et al. Synthesis of nitrogen and Sulfur Co-doped reduced graphene oxide as efficient metal-free cocatalyst for the photo-activity enhancement of CdS. Applied Catalysis B: Environmental. 2018;**236**:212-221. DOI: 10.1016/j.apcatb.2018.05.021

[6] Alam Z, Verma B, Sinha ASK. Creation of heterojunction in CdS supported on N, S-rGO for efficient charge separation for photo-reduction of water to hydrogen. International Journal of Hydrogen Energy. 2020;**45**: 4095-4112. DOI: 10.1016/j.ijhydene.2019.12.056

[7] Jingxiang L, Jiaguo Y, Mietek J, Swelm W, Al-G AA. Heterojunction photocatalysts. Advanced Materials. 2017;**29**:1601694. DOI: 10.1002/adma.201601694

[8] Amy LL, Guangquan L, John TY, Jr. Photocatalysis on TiO_2 surfaces. Principles, mechanisms, and selected results. Chemical Reviews. 1995;**95**: 735-758. DOI: 10.1021/cr00035a013

[9] Xin L, Rongchen S, Song M, Xiaobo C, Jun X. Graphene-based heterojunction photocatalysts. Applied Surface Science. 2018;**430**:53-107. DOI: 10.1016/j.apsusc.2017.08.194

[10] Huanli W, Lisha Z, Zhigang C, Junqing H, Shijie L, Wan Z. Jianshe Li, and Xinchen W. semiconductor heterojunction photocatalysts: Design, construction, and photocatalytic performances. Chemical Society Reviews. 2014;**43**:5234-5244. DOI: 10.1039/C4CS00126E

[11] Shao M, Yangfan S, Shengjie D, Rui T, Xiongwei Z, Lingmin Y. Weng Fai Ip Baomin X Xing-Qiang S, Yi-Yang S, Xuesen W, and Hui Pan. Carbonized MoS_2: Super-active Co-catalyst for highly efficient water splitting on CdS: ACS sustainable. Chemical Engineer. 2019;**7**:4220-4229. DOI: 10.1021/acssuschemeng.8b05917

[12] Guancai X, Liming G, Linjuan Z, Beidou G, Aisha B, Qi X, et al. Interaction-dependent interfacial charge-transfer behavior in solar water-splitting system. Nano Letters. 2019;**19**: 1234-1241. DOI: 10.1021/acs.nanolett.8b04768

[13] Guoning L, Charles K, Rong J, Shaopeng Q, Yongbing L, Jinxi C, et al.

MoS_2-stratified CdS-Cu_{2-x}S Core−Shell nanorods for highly efficient photocatalytic hydrogen production. ACS Nano. 2020;**14**:5468-5479. DOI: 10.1021/acsnano.9b09470

[14] Jing D, Wenjian F, Hui Y, Weiwei X, Xianghua Z, Wenfeng S. Few-layered MoS_2/ZnCdS/ZnS heterostructures with an enhanced photocatalytic hydrogen evolution. ACS Applied Energy Materials. 2022;**5**(4):4893-4902. DOI: 10.1021/acsaem.2c00301

[15] Menglong W, Zhixiao Q, Zhidan D, Rui L, Junbo Z, Dongmei M, et al. Red phosphorus/carbon nitride van der Waals heterostructure for photocatalytic pure water splitting under wide-Spectrum light irradiation: ACS sustainable. Chemical Engineer. 2020;**8**: 13459-13466. DOI: 10.1021/acssuschemeng.0c04372

[16] Kun C, Zongwei M, Tao W, Qing K, Shuxin Q, Jinhua Y. MoS_2/graphene cocatalyst for efficient photocatalytic H2 evolution under visible light irradiation. ACS Nano. 2014;**8**:7078-7087. DOI: 10.1021/nn5019945

[17] Zheng P, Qinggang Z, Tianyou P, Xiaohu Z. One-pot synthesis of reduced graphene oxide–cadmium sulfide nanocomposite and its photocatalytic hydrogen production. Physical Chemistry Chemical Physics. 2011;**13**: 21496-21502. DOI: 10.1039/C1CP22059D

[18] Quanjun X, Jiaguo Y, Mietek J. Synergetic effect of MoS_2 and graphene as cocatalysts for enhanced photocatalytic H_2 production activity of TiO_2 nanoparticles. Journal of the American Chemical Society. 2012; **134**(15):6575-6578. DOI: 10.1021/ja302846n

[19] Wolff CM, Frischmann PD, Schulze M, Bohn BJ, Wein R, Livadas P, et al. All-in-one visible- light-driven water splitting by combining nanoparticulate and molecular co-catalysts on CdS nanorods. Nature Energy. 2018;**3**:862-869 https://www.nature.com/articles/s41560-018-0229-6

[20] Ma D, Jian-W S, Yajun Z, Zhaoyang F, Xin J, C N. Highly efficient photocatalyst based on a CdS quantum dots/ZnO nanosheets 0D/2D heterojunction for hydrogen evolution from water splitting. ACS Applied Materials & Interfaces. 2017;**9**: 25377-25386 https://pubs.acs.org/doi/10.1021/acsami.7b08407

[21] Reshma B, Sarika K, Golu P, Rohan F, Dushyant K, Satishchandra O. $NiS_{1.97}$: A new efficient water oxidation catalyst for photoelectrochemical hydrogen generation. ACS Applied Materials & Interfaces. 2015;7: 20053-20060 https://pubs.acs.org/doi/10.1021/acsami.5b05077

[22] Amtul N, Satyabadi M, Parida KM. Heterojunction conception of n-$La_2Ti_2O_7$/p-CuO in the limelight of photocatalytic formation of hydrogen under visible light. RSC Advances. 2014; **4**:14633-14643 https://pubs.rsc.org/en/content/articlelanding/2014/ra/c3ra47037g

[23] Swarnava N, Takashi H, Sun S, Masao K, Tsutomu M, Kazunari D. Effects of Se incorporation in $La_5Ti_2CuS_5O_7$ by annealing on physical properties and photocatalytic H_2 evolution activity. ACS Applied Materials & Interfaces. 2019;**11**: 5595-5601 https://pubs.acs.org/doi/10.1021/acsami.8b02909

[24] Barawi M, Ferrer IJ, Ares JR, Sanchez C. Hydrogen evolution using palladium Sulfide (PdS) Nanocorals as photoanodes in aqueous solution. ACS Applied Materials & Interfaces. 2014;**6**:

20544-20549 https://pubs.acs.org/doi/10.1021/am5061504

[25] Zhenjiang L, Xuehua W, Wenli T, Alan M, Lina Y. ACS sustainable CoNi bimetal cocatalyst modifying a hierarchical $ZnIn_2S_4$ nanosheet-based microsphere Noble-metal-free photocatalyst for efficient visible-light-driven photocatalytic hydrogen production. Chemical Engineer. 2019;**7**: 20190-20201 https://pubs.acs.org/doi/10.1021/acssuschemeng.9b06430

[26] Min W, Peng J, Jiajia L, Yun Z, Xiuxun H, Zhaomin H. Facile synthesis of MoS_2/g-C_3N_4/GO ternary heterojunction with enhanced photocatalytic activity for water splitting. ACS Sustainable Chemistry & Engineering. 2017;**5**:7878-7886 https://pubs.acs.org/doi/10.1021/acssuschemeng.7b01386

[27] Haruki N, Tsuyoshi O, Seiji K, Morio N. Visible-light overall water splitting by CdS/WO_3/$CdWO_4$ Tricomposite photocatalyst suppressing Photocorrosion. ACS Applied Energy Materials. 2018;**1**:6730-6735 https://pubs.acs.org/doi/10.1021/acsaem.8b01600

[28] Hiromu K, Ryosuke A, Kosaku K, Akira Y, Masato K, Hideki K. Utilization of perovskite-type oxynitride $La_{0.5}Sr_{0.5}Ta_{0.5}Ti_{0.5}O_2N$ as an O_2-evolving photocatalyst in Z-scheme water splitting. ACS Applied Energy Materials. 2021;**4**:2056-2060 https://pubs.acs.org/doi/10.1021/acsaem.0c03055

[29] Shunya Y, Akihide I, Yun HN, Rose A, Akihiko K. Z-schematic solar water splitting using fine particles of H2-evolving $(CuGa)_{0.5}ZnS_2$ photocatalyst prepared by a flux method with chloride salts. ACS Applied Energy Materials. 2020;**3**:5684-5692 https://pubs.acs.org/doi/10.1021/acsaem.0c00661

[30] Katsuya I, Akihide I, Yun HN, Rose A, Akihiko K. Z-schematic water splitting into H_2 and O_2 using metal Sulfide as a hydrogen-evolving photocatalyst and reduced graphene oxide as a solid-state Electron mediator. Journal of the American Chemical Society. 2015;**137**:604-607 https://pubs.acs.org/doi/10.1021/ja511615s

[31] Yosuke YK, Yui G, Hikaru M, Hiroto I, Hajime S, Ryu A, et al. Z-scheme overall water splitting using ZnxCd1–xSe particles coated with metal cyanoferrates as hydrogen evolution photocatalysts. ACS Catalysis. 2021;**11**: 8004-8014 https://pubs.acs.org/doi/10.1021/acscatal.1c01187

[32] Parnapalle R, Dinesh KK, Dinesh S, Deepakkumar T, Vempuluru NR, Shankar MV, et al. Temperature-driven morphology control on CdSe nanofractals and its influence over the augmented rate of H_2 evolution: Charge separation via the S-scheme mechanism with incorporated Cu_3P. ACS Applied Energy Materials. 2021;**4**:13983-13996 https://pubs.acs.org/doi/10.1021/acsaem.1c02790

[33] Sooho L, Kangha L, Whi DK, Seokwon L, Do Joong S, Doh CL. Temperature-driven morphology control on CdSe nanofractals and its influence over the augmented rate of H_2 evolution: Charge separation via the S-scheme mechanism with incorporated Cu_3P. Journal of Physical Chemistry C. 2014; **118**:23627-23634 https://pubs.acs.org/doi/10.1021/jp508315m

[34] Deqian Z, Lang X, Wee-J O, Pengyuan W, Hongfei Z, Yuanzhi C, et al. Hierarchical $ZnIn_2S_4$/$MoSe_2$ nanoarchitectures for efficient Noble-metal-free photocatalytic hydrogen evolution under visible light. ChemSusChem. 2017;**10**:4624 https://chemistryeurope.onlinelibrary.wiley.com/doi/full/10.1002/cssc.201701345

[35] Xixiong J, Xiangqian F, Jianjian T, Ruolin C, Mengli L, Lingxia Z. MoS_2 quantum dot decorated g-C_3N_4 composite photocatalyst with enhanced hydrogen evolution performance. RSC Advances. 2016;**6**:52611-52619 https://pubs.rsc.org/en/content/articlelanding/2016/RA/C6RA07060D

[36] Nur F, Muhammad T. A critical review in strategies to improve photocatalytic water splitting towards hydrogen production. International Journal of Hydrogen Energy. 2019;**2**(44):540-577. DOI: 10.1016/j.ijhydene.2018.10.200

[37] Arora MK, Sahu N, Upadhyay SN, Sinha AS. Activity of cadmium Sulfide photocatalysts for hydrogen production from water: Role of support. Industrial and Engineering Chemistry Research. 1999;**38**:2659-2665. DOI: 10.1021/ie980400j

[38] John RH, Aarti M, Suljo L. Design principles for efficient and stable water splitting Photoelectrocatalysts. Accounts of Chemical Research. 2021;**54**:1992-2002 https://pubs.acs.org/doi/10.1021/acs.accounts.1c00072

[39] Fujishima A, Honda K. Electrochemical photolysis of water at a semiconductor electrode. Nature. 1972;**238**:37-38

[40] Hao W, Hui LT, Cui YT, Jason S, Lianzhou W, Rose A, et al. Photocatalytic and photoelectrochemical systems: Similarities and differences. Advanced Materials. 2019;**32**:1904717. Available from: https://onlinelibrary.wiley.com/doi/abs/10.1002/adma.201904717

[41] Amtul N, Kulamani P. n-$La_2Ti_2O_7$/p-$LaCrO_3$: A novel heterojunction-based composite photocatalyst with enhanced photoactivity towards hydrogen production. Journal of Materials Chemistry A. 2014;**2**:18405-18412 https://pubs.rsc.org/en/content/articlelanding/2014/ta/c4ta02401j

[42] Kousik B, Moumita C, Santimoy K, Debabrata P. Bimetallic PtAu alloy nanoparticles-integrated g-C_3N_4 hybrid as an efficient photocatalyst for water-to-hydrogen conversion. ACS Applied Materials & Interfaces. 2019;**11**:478-488 https://pubs.acs.org/doi/10.1021/acsami.8b121833

[43] Li Z, Wang X, Tian W, Meng A, Yang L. Sustainable CoNi bimetal cocatalyst modifying a hierarchical $ZnIn_2S_4$ nanosheet-based microsphere Noble-metal-free photocatalyst for efficient visible-light-driven photocatalytic hydrogen production. ACS Sustainable Chemistry & Engineering. 2019;**7**:20190-20201 https://pubs.acs.org/doi/10.1021/acssuschemeng.9b06430

[44] Wang M, Peng J, Li J, Zhao Y, Han X, Hao Z. Facile synthesis of MoS_2/g-C_3N_4/GO ternary heterojunction with enhanced photocatalytic activity for water. ACS Sustainable Chemistry & Engineering. 2017;**5**:7878-7886 https://pubs.acs.org/doi/10.1021/acssuschemeng.7b01386

[45] Yosuke K, Yui G, Hikaru M, Hiroto I, Hajime S, Ryu A, et al. Z-scheme overall water splitting using $Zn_xCd_{1-x}Se$ particles coated with metal cyanoferrates as hydrogen evolution photocatalyst. ACS Catalysis. 2021;**11**:8004-8014 https://pubs.acs.org/doi/10.1021/acscatal.1c01187

[46] Sooho L, Kangha L, Whi DK, Seokwon DJS, Doh CLJ. Its influence over the augmented rate of H_2 evolution: Charge separation via the S-scheme mechanism with incorporated Cu_3P. Physical Chemistry C. 2014;**118**:

23627-23634 https://pubs.acs.org/doi/10.1021/jp508315m

[47] Subhash CS, Santimoy K, Indranil M, Debabrata P, Prof. Dr. J M. Hierarchical $ZnIn_2S_4/MoSe_2$ nanoarchitectures for efficient Noble-metal-free photocatalytic hydrogen evolution under visible light. Chemistry - A European Journal. 2017;**23**: 14827 https://chemistry-europe.onlinelibrary.wiley.com/doi/10.1002/chem.201702561

[48] Singh SV, Gupta U, Biring S, Mukherjee B, Pal BN. In-situ grown nanoscale p-n heterojunction of Cu2S-TiO2 thin film for efficient photoelectrocatalytic H2 evolution. Surfaces and Interfaces. 2022;**28**:101660 https://www.sciencedirect.com/science/article/pii/S246802302100732X?via%3Dihub

[49] Wei Z, Yuanfu R, Yuying Z, Anqiang P, Ting Z. In-situ copper doping with ZnO/ZnS heterostructures to promote interfacial photocatalysis of microsized particles. ChemCatChem. 2021;**13**:564 https://chemistryeurope.onlinelibrary.wiley.com/doi/10.1002/cctc.202001407

[50] Mohamad FRS, Habib U, Robabeh B, Norani MM, Suriati S, Ng YH. Experimental and DFT insights on microflower $g\text{-}C_3N_4/BiVO_4$ photocatalyst for enhanced photoelectrochemical hydrogen generation from Lake water. ACS sustainable. Chemical Engineer. 2020;**8**: 9393-9403 https://pubs.acs.org/doi/10.1021/acssuschemeng.0c02063

[51] Yuguang C, Peng Z, Li N, Jianping L, Yong Y, Yelong Z, et al. Ultrathin visible-light-driven Mo incorporating In_2O_3–$ZnIn_2Se_4$ Z-scheme nanosheet photocatalysts. Advanced Materials. 2019;**31**:31. Available from: https://onlinelibrary.wiley.com/doi/10.1002/adma.201807226

Chapter 4

Conversion of Gas Turbine Combustors to Operate with a Hydrogen-Air Mixture: Modifications and Pollutant Emission Analysis

Rafael Domingues and Francisco Brójo

Abstract

In this work, an overview of the use of hydrogen in aviation, the modifications needed to adapt an existent gas turbine to use hydrogen, and a CFD simulation of an existent gas turbine burning hydrogen are performed. The CFD simulation was done in a CFM56-3 combustor burning hydrogen and Jet A. It was intended to evaluate the viability of conversion of existent gas turbines to hydrogen, in a combustion point of view, by analyzing the emissions while burning it through ICAO's LTO cycle. The pollutant emissions (only NOx, since hydrogen combustion produce only water vapor and NOx) were evaluated through a detailed mechanism and the Ansys Fluent NOx model to get a better agreement with the ICAO's values. For this assessment, several sensibility studies were made for hydrogen burn, for example, the analysis of the air flow with/without swirl in the primary zone and different inlet temperature and pressure for fuel. In the end, it was concluded that theoretically the CFM56-3 combustor can be converted to operate with hydrogen fuel with minor changes (related to injection system). The quantity of NOx produced for each power setting when burning hydrogen is expected to be almost twice the values for Jet A.

Keywords: CFM56-3, combustion chamber, pollutant emissions, jet fuel, hydrogen fuel

1. Introduction

The sustainable growth of aviation is important for the future of the economic growth, development, commerce, cultural exchange, and many other factors. According to some experts, by 2045, international air traffic is expected to increase by 3.3 times [1]. In 2015, international aviation consumed approximately 160 megatons (Mt) of fuel. By 2045, compared with the anticipated increase of 3.3 times growth in international air traffic, fuel consumption is projected to increase by 2.2–3.1 times compared with 2015, depending on the advances in technology and the Air Traffic Management (ATM) scenario [1].

The emissions resulting from the combustion of fossil fuels are usually considered as the main responsible for Greenhouse Gas (GHG) emissions, which are pointed as the primary factor that leads to global warming. For climate change, the primary concerns are emissions of CO, CO_2, NOx, and nvPM [2]. Also of concern are persistent contrails, which lead to cirrus clouds. Generally, it is the combination of a number of factors that determine the overall impact of the emissions on global surface temperature over a given timescale. These factors consist of quantities emitted, residence time, radiative forcing, and the temperature response profile of a particular pollutant [2].

The CO_2 emissions are of particular concern because of its exceptionally long residence time (thousands of years). Aviation today accounts for 2–3% of global CO_2 emissions. While at the global level, CO_2 emissions are increasing by around 3% per year, aviation's emissions covered by the EU ETS have increased on average by 5% year-on-year between 2013 and 2018. By 2040, it is expected that international aviation emissions could rise by up to 150% compared with 2020. These growth forecasts take into account the incremental technology improvements that may reduce fuel consumption and emissions by 1–1.5% annually [3].

About the NOx emissions, they are evaluated in two possible scenarios, which are landing and take-off (LTO) NOx emissions, which primarily affect local air quality, and full-flight NOx emissions, which have more effect on the global climate. In 2015, LTO NOx emissions were approximately 0.18 Mt., and by 2045, they are projected to range from 0.44 to 0.80 Mt. depending on the technology and ATM scenario [1]. While, in 2015, the full-flight NOx emissions of international aviation were 2.50 Mt., by 2045, the full-flight NOx emission projection ranges from 5.53 to 8.16 Mt., which represents a 2.2–3.3 times growth compared with 2015 [1].

To mitigate this problem, there are several possible solutions. On the one hand, the fuel burn reductions through the upgrade of the technology employed in the actual aircrafts such as the airframes (aerodynamics and mass) and the engines, both with the aim of achieve higher efficiency [2]; on the other hand, the use of alternative fuels and power sources [4, 5]. According to the ICAO 2016 trends assessment, a 100% substitution of aviation fuel with SAF could reduce 63% of the baseline CO_2 emissions from international flights in 2050 [4]. As referred by ATAG [5], it is possible that aviation meets net-zero CO_2 emissions by 2050; however, it would take an enormous effort to make it a reality. This would mean a rapid and massive transformation on aviation's energy supply through the use of SAFs, and it would require acceleration in aircraft and engine technology development, mainly: electric-, hybrid-, and hydrogen-powered aircraft.

Within this context, the conversion of the current gas turbine engines to new sustainable fuels can also be a solution. So, in this study, be analyzed the feasibility of the use of hydrogen fuel as substitute of the conventional jet fuel in a CFM56–3 combustor using a CFD approach. The NOx emissions produced while completing the ICAO's LTO cycle burning this fuel will be assessed for the standard operating conditions of the engine, as well as the influence of several operating parameters (swirl effect, temperature, and pressure of fuel) in these emissions.

1.1 Brief historical review

To date, the largest user of hydrogen in aeronautics is the space program where it is used as fuel for the rocket engines of launch vehicles. The first successful launch of a space vehicle propelled by a liquid hydrogen/liquid oxygen rocket engine took place at

Cape Kennedy on November 27, 1963. Several other rocket engine manufacturers in the United States were involved in the development of designs using LH_2; for example, the General Electric Company, the Rocketdyne Division of North American Aviation (now Rockwell International, Inc.), and the Aerojet General Corporation were among the leaders. Of the designs developed by these companies, the Rocketdyne J2 engine is an example, which has been eminently successful. It was used in both the second and third stages of the Saturn V launch vehicle, used in the Apollo program, which landed U.S. astronauts on the moon. In all of the launches of the Apollo program, there has never been a failure of one of the hydrogen-fueled rocket engines.

However, the space applications are relatively recent, if we look at history, the first reported use of hydrogen in aeronautics was a long time ago. According to Brewer [6], hydrogen was first employed as lifting medium when, in France, a small silk balloon was constructed by the Roberts brothers, under the direction of physicist J.A.C. Charles, and it was flown in Paris on August 27, 1783. This balloon rose to a height of 3000 ft. (914.4 m) and traveled a distance of 15mi (24.14 km). In that year, on December 1, a larger hydrogen-filled model, which carried two passengers, the physicist Charles and one of the Roberts brothers, was launched. This flight traveled 25mi (40 km) from Paris in less than 2 hours.

Later in history, airships came into being as a result of man's desire to control the direction and speed of flight. Numerous attempts were made to achieve such control with balloons without measurable success until 1852, when a Frenchman, Henri Giffard, constructed an airship on which he mounted a steam engine of his own design. Giffard flew this hydrogen-filled airship from the Hippodrome in Paris on September 24, 1852, attained an estimated speed of 6 mph, and demonstrated the first appreciable control of a "lighter-than-air craft." In 1872, Paul Haenlein developed and flew an airship powered by an internal combustion engine, which was fueled by gaseous hydrogen that was drawn from the lifting cells of the airship envelope [6]. A significant step leading to the use of hydrogen in commercial air transportation occurred in 1900 when the first rigid airship designed by Count Ferdinand von Zeppelin, the LZ-1, made a successful flight. In 1911, commercial air operations were started by a German transportation company (DELAG), using five Zeppelin airships. In October 1924, the Zeppelin factory at Lake Constance, in Germany, completed the construction of the LZ-126, inflated it with hydrogen, and delivered it to the United States by a transatlantic flight.

In 1955, a report by Silverstein and Hall of the (then) NACA-Lewis Flight Propulsion Laboratory was published in which the potential of liquid hydrogen as a fuel for use in both subsonic and supersonic aircraft was explored. According to it, theoretically, the use of hydrogen fuel could significantly improve the maximum range [6]. As a result of this study, an experimental program with a U.S. Air Force B-57 twin-engine medium bomber was initiated to demonstrate the feasibility of burning hydrogen in a turbojet engine at high altitude. The modified aircraft was first flown in 1956.

From 1954 to 1955, Lockheed Aircraft Corporation made a series of conceptual design studies of hydrogen-fueled aircraft in cooperation with Pratt & Whitney Aircraft and the Rex Division of AiResearch Corporation. In 1956, the U.S. Air Force awarded a contract to Lockheed's Advanced Development Projects organization to build two prototype aircrafts (known as CL-400), which would be capable of cruising at Mach 2.5 at 100000 ft. (30,480 m) altitude. This aircraft was to carry a two-man crew, and the main objective was related to long-range reconnaissance missions.

Also in 1956, at the same time the U.S. Air Force contracted with Lockheed for the development of the CL-400 airplane, and the Pratt & Whitney Aircraft Division of United Aircraft Corporation was awarded a contract to investigate the feasibility of using LH_2 as a fuel in aircraft engines. The work at Pratt & Whitney covered a broad spectrum ranging from applied research efforts such as heat transfer and materials investigations, to development testing of a J57 engine modified to operate on LH_2. It also included the design, construction, and test of a new design of engine (the Model 304). Conversion of the J57 to operate on LH_2 was accomplished in just 5 months, and the first tests were performed in the fall of 1956 [6]. The work with the J57 showed that conventional jet engines could be readily adapted to use LH_2 fuel. In this research, after examining many possible cycles, the Hydrogen Expander cycle (this cycle is well explained by Brewer [6]) was selected for experimental evaluation to create the Model 304 engine. This was a unique cycle developed specifically to take advantage of the properties of hydrogen and to meet the performance requirements of the CL-400 airplane. The first demonstration test of a complete 304 engine was accomplished in September 1957.

In spite of the success in developing practical solutions to the problems encountered with handling the cryogenic liquid fuel, the CL-400 aircraft was never built due to performance and logistics limitations. So, in 1957, the program was terminated by mutual agreement between the Air Force, Lockheed, and Pratt & Whitney. However, the CL-400 design and development program showed that it was entirely feasible to build a hydrogen-fueled airplane.

In the 1970s, Lockheed performed studies on different liquid hydrogen-fueled subsonic cargo and passenger transport jets for NASA Langley Research Center. The results are presented in the NASA-reports NASA CR-132558, NASA CR-132559, and NASA CR-144935. The main conclusions from these and furthers studies have been summarized by Daniel G. Brewer in [6]. The studies showed that hydrogen propulsion is especially beneficial in terms of energy use for long-range aircraft with internal hydrogen tanks.

In the 1980s, Tupolev developed the Tu-155 that was based on the medium-range transport aircraft TU-154B. Moreover, the TU-155 was built and successfully tested without any serious incidents, and it first flew burning hydrogen in one of its three engines in April 1988. The modified engine was also able to be run with natural gas. The TU-155 was followed by the TU-156 that could be run with natural gas or kerosene [7].

At the beginning of the twenty-first century, the Cryoplane Project comprised of 36 European research partners from industry, universities, and research institutions. They contributed to this project covering aircraft configuration, systems and components, propulsion, safety, environmental compatibility, fuel sources and infrastructure, transition. The total project time was 26 months and started in April 2000. During this project, several conventional and unconventional overall aircraft design studies and detailed investigations of hydrogen fuel systems and components were performed [7, 8].

More recently, in July 2010, Boeing unveiled its hydrogen-powered Phantom Eye UAV that uses two converted Ford Motor Company piston engines. Nowadays, governments and companies are investing again in hydrogen's potential. For instance, the ENABLEH2 (ENABLing cryogEnic hydrogen-based CO_2 free air transport) consortium was given such a hand, almost 20 years after the European Commission's last attempt to ramp up LH_2 research and development under the Cryoplane project. The project's objective is to demonstrate that switching to hydrogen is feasible and must complement research and development into areas such as advanced airframes, propulsion systems, and air transport operations [9]. Another example is the project named ZEROe, announced by Airbus in September 2020, which has the ambition to develop the world's first zero-emission commercial aircraft. This project consists of the development of three concept planes (powered by hydrogen combustion through modified gas turbine engines or hybrid systems), which Airbus says that could be ready for deployment by 2035 [10].

2. Basic engine principles

2.1 CFM56-3 general specifications

The CFM56-3 is a high bypass, dual-rotor (or dual-shaft), axial flow turbofan engine, and this particular variant of CFM56-3 has a bypass ratio of 5.1:1 and a dry weight of 1966 kg [11, 12]. Its dual-shaft design consists of a fan and booster (low-pressure compressor), high-pressure compressor, annular combustion chamber, and a high- and low-pressure turbine section. The two shafts respectively connect the low- and high-pressure sections using a five-bearing system (three roller, and two ball bearings) [13].

2.2 Combustor

First of all, it is important to understand the difference between the combustor and the combustion chamber [14]. The combustor includes all of the combustion systems, that is, the diffuser, the combustion chamber, the inner and outer casing, the spark plugs, and the fuel injectors, whereas the combustion chamber refers to the exact place in which combustion takes place.

The main purpose of a gas turbine combustor is to introduce heat energy into the mass of air previously compressed (in the compressor) [15], by burning fuel in it so that the products of combustion can be expanded to get useful work output (absorbed by the turbines) and then, on their discharge to atmosphere, provide a propulsive jet [16]. Due to space limitations and requirements of energy and momentum, the volume flow rate as well as rate of heat release is very high in a gas turbine combustion chamber and the residence time of fuel is very small, of the order of a few milliseconds [15]. In gas turbines, the combustion is a continuous process that takes place at high pressure in a smaller space and usually at a very high temperature [16]. Thus, continuously high combustion temperatures, large continuous flow, and high heat energy release make the design and development of a gas turbine combustor rather difficult [15].

A gas turbine combustor must satisfy a wide range of requirements. However, for the aircrafts, the priorities are the reliability, the low fuel consumption, low pollutant emissions, engine size, and weight [17].

The choice of a particular combustor type and design is determined largely by the overall engine design and by the need to use the available space as effectively as possible [16]. Overall, the combustors may be subdivided into three main types: tubular or can, tubo-annular, and annular [16]. The CFM56-3 has an annular combustor, and **Figure 1** shows it during the digitalization process to obtain the model of geometry used in this work.

The annular configuration is used by most modern jet engines because of its lighter design. This type of combustor represents the ideal configuration for combustors since

Figure 1.
CFM56-3 combustor photograph [14].

its "clean" aerodynamic layout results in compact dimensions [16]. This configuration consists of an annular liner that is mounted concentrically inside an annular casing. Among the advantages, the annular combustion chambers have the least pressure drop due to larger volume per unit surface area and are more efficient than can-type chambers. It also requires about half of the diameter of can-type chambers for the same mass flow [15].

Its main drawback stems from the heavy buckling load on the outer liner [17]. Moreover, any change in the flow velocity profile can result in significant change in the temperature distribution of the outlet gases, and distortion of inner annular chamber is critical because it disrupts the flow of cooling air and also changes the outlet temperature distribution. This is because of lower degree of curvature of the chamber surfaces [15]. Another drawback is related to experimental tests of this type of combustion chamber. At full-load conditions, the tests of large annular combustion chambers supplying air at the levels of pressure, temperature, and flow rate required are extremely difficult, and the cost is very high [17]. Nowadays, there are very few facilities worldwide that can supply air in those conditions [18].

3. Fuels in gas turbine engines

At the current time, almost all aviation fuel (jet fuel) is extracted from the middle distillates of crude oil (kerosene fraction), which distils between the gasoline and the diesel fractions [18]. The Jet A and Jet A-1 grades are the most used kerosene-type fuels worldwide in civil aviation.

As referred before, to mitigate the problem of the pollutant emissions in the future, there seem to be two viable solutions, the use of SAFs or the use of hydrogen fuel (through hydrogen combustion engines or fuel cells). In this work, will only be considered the hydrogen for the combustion gas turbine engines (GTEs).

3.1 Hydrogen

The hydrogen is the simplest and most abundant of the chemical elements in the universe. On Earth, under standard conditions (H_2), its concentration is negligible. However, in chemically combined form, it is the third most abundant element on

Earth. Consequently, hydrogen can be considered an energy carrier, similar to electricity, since it can be produced from other compounds, mostly from hydrocarbons and water [19]. The main reasons why hydrogen can play an increasingly significant role in meeting the worlds energy demands and addressing environmental concerns are that hydrogen meets three important criteria: a promising low-carbon alternative reducing emissions of GHG, providing energy security, and the possibility of reducing local pollutants, that is, NOx and particulates [19].

On a commercial scale, despite the several production sources, most of the hydrogen is currently produced through the process of steam reforming of methane (SRM) from natural gas [19]. As part of the energy future, the various hydrogen sources can be grouped into three types, namely fossil fuels (coal, natural gas, petroleum, oil shale, etc.), renewable sources (biofuels, water, photovoltaic, solar, algae, etc.), and nuclear (e.g., using thermal energy from nuclear reactions for water splitting). For hydrogen to be part of a sustainable energy future, renewable and nuclear sources need to play a more significant role in hydrogen production, and cost-effective carbon capture and storage technologies need to be developed and upgraded.

3.2 Utilization of hydrogen in gas turbine engines

For the use of the hydrogen fuel in the current GTEs, in theory, the minimum modifications needed are the change of the injection system and the implementation of facilities to evaporate the hydrogen, which is stored in the tanks in a liquid (or cryogenic) state. This can be accomplished by an external heat source or a heat exchanger (HE) [20]. However, to take full advantage of the hydrogen's distinct thermo-transport properties (high diffusivity, low ignition energy, wide flammability limits, and the highest laminar flame speed) that make its combustion and emission characteristics notably different from those of hydrocarbons fuels, beyond the injection system, the combustion chamber must be also changed [19]. In **Table 1**, is provided a comparison between the main properties of hydrogen and Jet A [6, 9].

When changing to hydrogen, either the combustor outlet temperature (COT) or the net thrust could be retained [20]. Because of the considerably higher heating value of hydrogen, the fuel flow to achieve the same COT or net thrust is reduced by almost

Properties	**Units**	**H2**	**Jet A**
Liquid density	$g\ cm^{-3}$(at 283 k)	0.071	~0.811
Melting point	K	14.01	−263
Boiling point	K (at 1 atm)	20.27	440–539
Heat of vaporization	$J\ g^{-1}$ (at 1 atm)	446	360
Specific heat	$J\ g^{-1}$	9.69	1.98
Lower heating value	$MJ\ kg^{-1}$	119.96	43.15
Flammability limits in air	vol%	4.0–75.0	0.6–4.7
Thermal energy radiated to surrounding	%	17–25	30–42
Diffusion velocity in NTP air	$cm\ s^{-1}$	≤2.00	<0.17
Flame temperature in air	K (stoichiometric)	2318	2200

Table 1.
Properties for hydrogen fuel and Jet A fuel.

two-thirds. When the COT is preserved, the net thrust increases, resulting in a corresponding increased specific thrust. When it is opted to retain the net thrust, this results in a lower COT [20]. According to Boggia and Jackson [21], the performance improvements could be explained by two fundamental changes when using hydrogen: reduced mass flow and changed composition of the gases expanding through the turbine(s). While the latter improves the performance, the former deteriorates the performance. Reduced mass flow through the turbine lowers the thrust output for two reasons. First, decreasing the fuel flow implies that the exhaust mass flow decreases accordingly; hence, without any variation in gas composition, the thrust output decreases. In second, a reduced mass flow through the turbine will result in a higher total temperature drop and, thereby, also a higher total pressure drop across the turbine in order to deliver the same amount of power to the compressor. Because of the lower total temperature and pressure at the turbine exit, both the pressure thrust (thrust due to different pressure at engine inlet and exit) and momentum thrust decrease (the effect of decreased core nozzle velocity).

However, the loss in thrust due to reduced mass flow is offset by the increased thrust owing to changed properties of the combustion products [21]. With the use of hydrogen, the combustion products contain no CO_2 and a larger portion of H_2O, which has a higher C_p value than CO_2. Having investigated a simple turbojet engine, Boggia and Jackson [21] concluded that the C_p value has increased by ~4% in the hot section of the engine when changing to hydrogen fuel. Increased C_p value through the turbine will similarly, but in the opposite direction as reduced mass flow, affect the performance. For a fixed power output, it will cause smaller total temperature and pressure drops across the turbine. Provided that the core nozzle is not choked, a larger nozzle expansion ratio will result in a larger exhaust velocity, which, in turn, will increase the momentum thrust. In total, the positive effect of increased C_p value outweighs the effect of reduced mass flow and, hence, results in an increased net thrust when switching to hydrogen and retaining the COT [20].

It should be pointed out that the energy consumption to attain a certain COT is highly dependent on the fuel-injection temperature and the location of the heat exchanger (HE) used to evaporate the liquid hydrogen. By heating the fuel more, it is possible to achieve performance benefits. The effects on engine performance are quite small, but still there are some desired features that could be exploited [20].

If the COT is kept the same, the turbine entry temperature (TET) is also about the same, thus requiring the same cooling technology [21]. On the other hand, the option of lowering the COT to preserve the net thrust will lead to a decrease in TET. So, this will require less advanced cooling technology as well as having a favorable effect on turbine blade life. Moreover, designing for a lower maximum cycle temperature will help to suppress the NOx emissions.

3.3 Mechanisms of pollutant formation

The concentration level of pollutants in gas turbine exhaust can be related directly to several factors that control the emissions in conventional combustors. These factors may be considered in terms of primary-zone temperature, equivalence ratio, degree of homogeneity of the primary-zone combustion process, residence time in the primary zone, liner-wall quenching characteristics, and fuel spray characteristics [17]. These factors vary from one combustor to another and, for any given combustor, with changes in operating conditions [17].

For the conventional fuels, such as hydrocarbons (in this case Jet A) or even the SAFs, the pollutant emissions of most concern are CO, CO_2, UHC, NOx, and PM (or smoke). From the environmental standpoint, hydrogen is nearly a clean fuel once it produces only NOx (considering that water vapor is not a major pollutant) [19]. So, in this work, only the NOx emissions will be presented.

About the NOx (NO plus NO_2 emissions), in conventional gas turbine combustors, there are four main mechanisms that are responsible for the NOx formation: thermal NO, nitrous oxide (N_2O) mechanism, prompt NO, and fuel NO. The last one is usually of less importance for normal fuels (there is no fuel-bond nitrogen) [17]. In the case of hydrogen burn, we must still consider the NO formation through intermediate NO_2 [19].

For hydrocarbon fuels, the two main mechanisms that are responsible for the formation of NOx are thermal NO and prompt NO, while for hydrogen flames the two main mechanisms associated with NOx formation are the thermal NO and NO formation through intermediate NO_2. So, in this subsection only these three mechanisms will be referred. For more information about the others, there are a good review of them in Lefebre et Ballal and Kenneth Kuo [17, 22].

The thermal NO is produced by the oxidation of atmospheric nitrogen (N_2) in high-temperature regions of the flame and in the post-flame gases [22]. This endothermic process is controlled largely by flame temperature, and it proceeds at a significant rate only at temperatures above around 1850 K (it requires the breaking of the tight N_2 bond) [17, 19, 22]. For the typical conditions encountered in conventional gas turbine combustors (high temperatures for only a few milliseconds), NO increases linearly with residence time, but does not attain its equilibrium value. The extended Zeldovich mechanism is utilized by the most of the proposed reaction schemes for thermal NO. The principal reactions of this mechanism are represented in Eqs. (1)–(4) [17, 19]:

$$O_2 \leftrightharpoons 2O \tag{1}$$

$$N_2 + O \leftrightharpoons NO + N \tag{2}$$

$$N + O_2 \leftrightharpoons NO + O \tag{3}$$

$$N + OH \leftrightharpoons NO + H \tag{4}$$

The prompt NO can be formed in a significant quantity in some combustion environments such as in low-temperature, fuel-rich conditions and when residence times are short. These conditions can be created in gas turbines [18].

In hydrocarbon flames, prompt NO occurs in the earliest stage of combustion and its formation is associated with the reaction of molecular N_2 with radicals, such as C, CH, and CH_2, which are fragments derived from fuel, through a complex series of reactions and many possible intermediate species. Some of these reactions are represented in Eqs. (5)–(9):

$$CH + N_2 \rightleftharpoons HCN + N \tag{5}$$

$$N + O_2 \rightleftharpoons NO + O \tag{6}$$

$$HCN + OH \rightleftharpoons CN + H_2O \tag{7}$$

$$CN + N_2 \rightleftharpoons NO + CO \tag{8}$$

$$CH_2 + N_2 \leftrightharpoons HCN + NH \tag{9}$$

For hydrocarbon flames, the major contribution is from CH and CH_2 species, as shown in Eqs. (5) and (9). The products of these reactions could lead to formation of amines and cyano compounds that subsequently can react with species such as N, O, or OH to form NO by reactions like those occurring in oxidation of fuel nitrogen or oxidation of other nitrogen species. At present, the prompt NO contribution to total NOx from stationary combustors is small. However, as NOx emissions are reduced to very low levels by employing new strategies that tend to reduce the flame temperature (such as burner design or geometry modification), the relative importance of the prompt NO can be expected to increase [18].

For hydrogen flames, the second mechanism that is relevant for the NOx emissions corresponds to the NO formation through intermediate NO_2. This mechanism may be represented by the **Eqs. (10)** and **(11)** [19]:

$$NO + HO_2 \rightarrow NO_2 + OH \tag{10}$$

$$NO_2 + H \rightarrow NO + OH \tag{11}$$

3.4 Influence of temperature/pressure in NOx formation with hydrogen fuel

Owing to its high adiabatic flame temperature, hydrogen combustion produces significant NOx. Therefore, by increasing the strain rate (a_s), which is pressure-dependent, the flame temperature can be lowered and NOx emissions are reduced. So, the pressure can be also a controlling parameter for NOx formation with hydrogen flames.

This way, the thermal NO mechanism is dominant at low pressures, whereas NO formation *via* intermediate NO_2 becomes important at moderate pressures once in that condition, the flame temperature decreases for a given strain rate, a_s, due to enhanced recombination by the reaction [19]:

$$\mathrm{H + O2 + M \rightarrow HO2 + M} \tag{12}$$

Consequently, the maximum NO formation decreases for moderate pressures [22].

At higher pressures, the net effect of reactions (10) and (11) is to $H + HO_2 \rightarrow 2OH$, through which radicals H, OH, and O are produced [19]. This enhances the formation of thermal NO. Some studies refer that under specific conditions, the formation of NO can be dominant over NO_2.

Still considering the influence of pressure, but now for hydrocarbon fuels, the N_2O and prompt mechanisms dominate at low temperature and are independent of pressure, whereas the higher NOx levels associated with higher combustion temperature are primarily due to thermal NO, which exhibits a square-root dependence on pressure [22].

3.5 Chemical kinetic mechanisms

Chemical kinetics is a capital point when modeling a combustion problem. In this case, the fuel combustion kinetics is extremely important in order to develop a model that allows a good emission prediction from the engine. Without proper kinetics all the attempts will go in vain. The development of detailed chemical kinetic models is extremely challenging once typical fuels (such as gasoline, diesel, and jet fuels) derived from different sources can be composed of hundreds to thousands of compounds. Therefore, detailed kinetic models for such fuels cannot contain all the compounds due to the limitation of current computational resources [23]. For that reason, a simplified mixture called surrogate mixture must be defined and used to develop a kinetic model.

For the emission predictions, the kinetic model development can be even more difficult once the NOx chemistry must be developed together with fuel chemistry making realistic chemistry getting even more complicated. In this section, will be introduced the Jet A and hydrogen combustion kinetic models applied in this study.

3.5.1 Jet A kinetic mechanisms

Although kinetic models of jet fuel are still underdeveloped, significant progress has been made in this area in the recent decades. Jet fuels are kerosene-type cut of petroleum containing C-10–C-18 hydrocarbons, including alkanes, cycloalkanes, and aromatic compounds. The literature reviews show that there are several kinetic models available for jet fuel combustion and some of these models are listed (and briefly descripted) by Mostafa [23].

However, only few of them are suitable to satisfy our current needs. Based on our objective to predict aircraft engine emission (specifically NOx) using CFD simulation, we need at first a jet fuel kinetic mechanism that allows to simulate the combustion process and, then, one that fairly predicts NOx formation in this combustion chamber. As CFD with kinetic models is computationally highly expensive, the number of species in the kinetic scheme needs to be limited. For that reason, after evaluating the kinetic models available, it was opted by the option presented by Kundu et al. [24].

Kundu et al. [24] proposed a simplified kinetic mechanism with NOx chemistry based on 17 species and 26-step reaction for Jet A (17 steps for Jet A reaction and nine steps for the sub-mechanism for NOx prediction). This mechanism has been developed specifically to predict NOx formation during combustion of aviation kerosene. However, the mechanism does not cover the entire range of pollutant species, once to limit the number of species, the mechanism does not include NO_2. Despite the limitation of this kinetic model, it was used in this work. This was not the best option, but among the available ones was the least bad.

3.5.2 Hydrogen kinetic mechanisms

The hydrogen oxidation chemistry represents the most fundamental and important building block in the hierarchy of hydrocarbon chemistry. Consequently, its chemistry has been extensively investigated, and a large number of detailed mechanisms (that can be found in the literature), including H_2/O_2 kinetics, have been developed and validated using different combustion configurations [19, 25]. Some of these mechanisms have been optimized for the combustion of pure hydrogen, but most of them are dedicated to the combustion of hydrocarbons including submechanisms for H_2/O_2 chemistry. However, the accuracy of the H_2/O_2 subset is also essential for the overall performance of a hydrocarbon mechanism [19].

To choose the kinetic model more adjusted for this case, some reviews available in the literature [25, 26] were analyzed. For instance, Ströhl et al. [25] made an evaluation of detailed reaction mechanisms for hydrogen combustion under gas turbine conditions. That study shows that the mechanisms of Li et al. or Ó Conaire accurately represent H_2/O_2 kinetics under gas turbine conditions. However, it suggests that the Li et al. mechanism is best suited for the prediction of H_2/O_2 chemistry since it includes more up-to-date data for the range of interest [25]. Also, the Li et al. mechanism has been found to provide the best match with measurements over a wide range of equivalence ratio and pressure, using various targets, including shock tube ignition delay and laminar flame speed data [26].

So, it was concluded that the Li et al. mechanism should perform better than the others, and for that reason it will be used in this work. For the NOx prediction (of hydrogen burn), was used the NOx sub-mechanism based on the study by the Glarborg group that is available in the database of ANSYS Fluent 2020R2 together with the mechanism of Li et al.

4. Materials and methods

4.1 Geometry

For this study, was used a CAD design of the CFM56–3 combustor made in the study [14], based on the CFM56-3 combustor of **Figure 1**. Due to the existent symmetry in the CFM56-3 combustor and in order to decrease the simulation time and effectively represent the four fuel injectors (in the 20) that supply a richer mixture, it will be used only a quarter section of the combustor for simulation purposes; that is, there will be one fuel injector for every five injectors present in each quarter section of the model combustor that supplies an even richer mixture. In this study, the rich fuel injector is the middle one (element 24 of the **Figure 2b**). All the details present in the combustor geometry are represented in **Figure 2**, including the combustor walls, dome, dilution holes, fuel injectors, and primary/secondary swirlers.

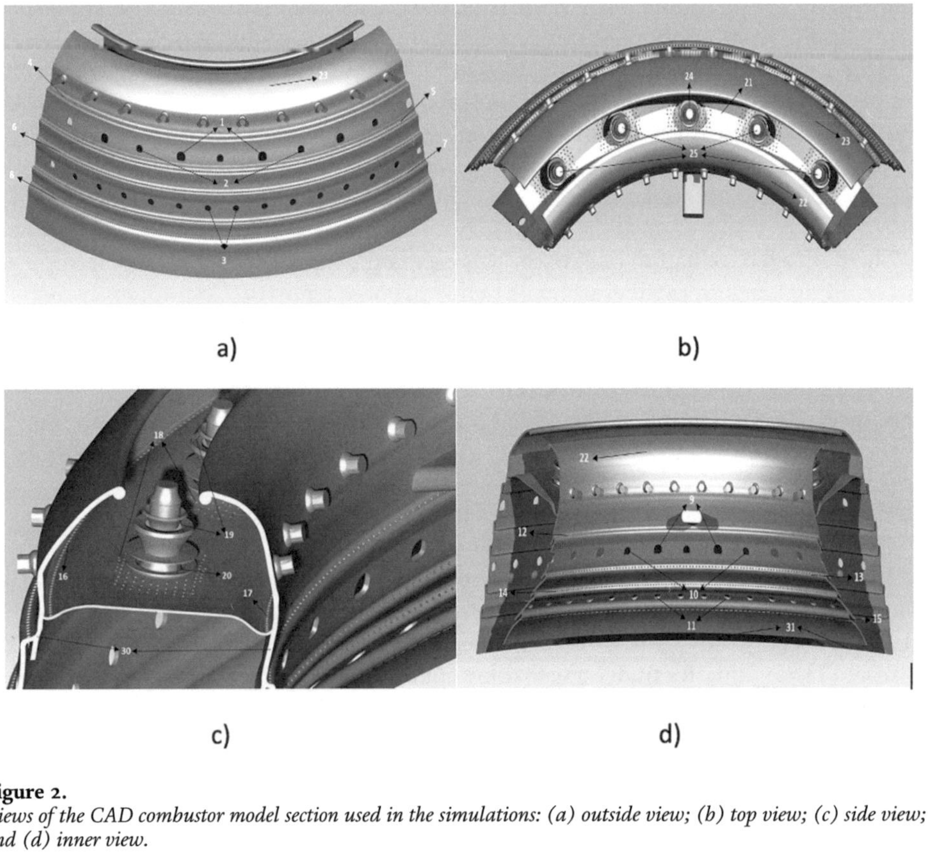

Figure 2.
Views of the CAD combustor model section used in the simulations: (a) outside view; (b) top view; (c) side view; and (d) inner view.

4.2 Modifications in the original geometry

As told by Oliveira [14], there were some parts of the combustor such as the swirlers that the exact geometry was not achieved. Furthermore, after analyzing some documentation of the CFM56-3 engine, photos of the combustor and doing the preliminary simulations, there were found some problems related with the fuel injectors position, the shape of the exit of the secondary swirlers, and the connection of this exits with the cooling walls of the dome. All of these problems were affecting the results. For that reason, some changes have been attempted to try to correct these small problems. All the changes in the model were performed in CATIA V5 R20. In **Figure 3** are presented two cut-view images of the first swirler, in which the first shows the original CAD model received and the second the final CAD model with the modifications made during this work.

During the early phases of the work, it was concluded that even so, the simulation time for the quarter section would be very large. So, to test new sets of modifications needed in the geometry, mesh, or the used models, a geometry where only one injector was represented was developed. Thereby, some sets of modifications could be excluded without spending the total time of the simulation in the quarter section.

4.3 Mesh generation

In this work, several meshing software (HELYX-OS, SnappyHexMesh, Simscale and Fluent Meshing) were tried in order to get the best mesh possible with the computational resources available (mainly the quantity of RAM). Due to the complexity of the geometry, the only software that provided a good quality mesh was the Fluent Meshing after using the set of tips provided by ANSYS in [27] about the best practices for gas turbine combustion meshing. This way, it was possible to create a good enough mesh for the simulation, where all the features of the geometry were correctly represented with the available computational resources.

After analyzing several meshes, the independency test was performed using three meshes, with coarse, medium, and fine refinement, having 11,830.638 cells, 16.318.327 cells, and 22.602.875 cells, respectively. The data collected relative to numerical/experimental data for Jet A fuel (no experimental data relative to H_2 combustion in this GTE was available) were presented by Ribeiro [28] and ICAO [12], and

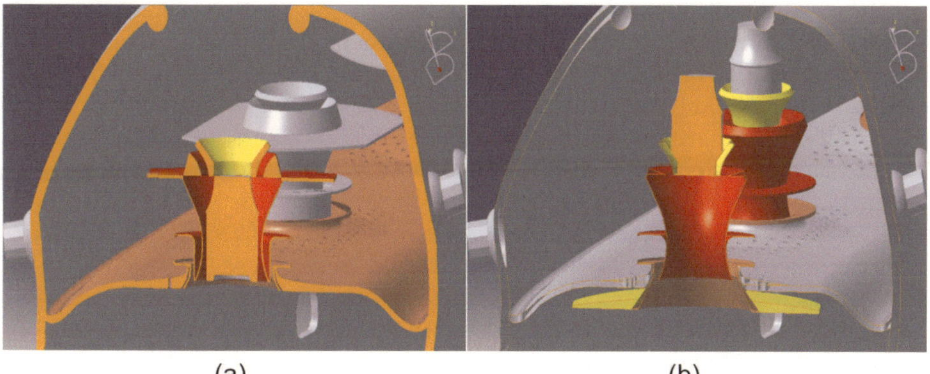

Figure 3.
Cut view of the models: (a) original model and (b) modified model.

these documents contain the exit temperature of the combustion chamber (only for the operating condition of 100%) and the NOx emissions of the engine, respectively. For that reason, the parameters used to analyze the independency of the mesh were:

- The average static temperature of the combustion chamber in the outlet;
- The average static temperature and velocity in a defined plane (cut-view of the first swirler), once the NOx emissions depend mainly of the temperature and residence time inside the combustion chamber.

All the independency tests were performed for the operating condition of 100%, since it is the only operating condition where the outlet temperature of the chamber is known. The maximum difference occurred for the average static temperature in the plane of the cut view of the swirler, and it was in the order of 2.16% between the values of the coarse and the intermediate meshes; however, between the values of the intermediate and the fine meshes, the difference was only 0.1%. The difference between the values of the average static temperatures for the outlet was nearly 0% between the coarse and the intermediate meshes as well as between the intermediate and the fine meshes. For the velocity magnitude, in the cut-view plane, the difference between the values was approximately 0.71%, between the values of the coarse and intermediate meshes, and 0.49%, between the values of the intermediate and the fine meshes.

4.4 Numeric simulation

The software used to perform this study was ANSYS Fluent 2020R2 [29]. Double-precision option was enabled, once a small error in this case can influence largely the results of the models. For the setup, the energy model was enabled. This model must be activated as this regards the energy related to the temperature change within the combustion process or heat transfer. For the viscous model, as was utilized a step-by-step solution, the first step was made with the realizable k-? and then was used the RSM in the other steps. For the radiation model, the P1 radiation model was chosen to simulate the heat transfer by radiation. This model was chosen for this study because it is accurate enough and reduces the computational cost in relation to the other models. Concerning the species, two models were used. At first to calculate an initial guess, the non-premixed combustion and species transport with one equation were used, for Jet A and hydrogen fuel simulations, respectively. And then, after obtaining a first initial solution converged (or almost converged), the detailed mechanisms referred before (Kundu et al. for Jet A and Li et al. for hydrogen) were imported, and the simulations were resumed until obtaining completely converged solutions. In this case, the used Turbulence-Chemistry Interaction was Eddy-Dissipation and the Chemistry Solver was the Relax to Chemical Equilibrium.

To evaluate the NOx emissions, two approaches were followed. The first one was to use the sub-mechanisms provided in the detailed mechanisms. These sub-mechanisms presented before can cause some problems once; for instance, the sub-mechanism for NOx produced by Jet A does not include the NO_2 species, which clearly will affect the results. The second approach was to use the NOx model provided in ANSYS Fluent. This model must be enabled to ANSYS Fluent display information regarding NOx formation during the solution calculation, or it can be calculated in post-processing (the approach chosen in this work). In the end, an assessment must be made to reveal which approach is in better agreement with the ICAO's database values (for the Jet A), to be used in the work.

The boundary conditions were defined through an iterative process that has three phases. In the first one, through an iterative process (simulation and result analysis) and the available data present by Ribeiro [28] regarding the conditions of each stage of the GTE (mainly air mass flow, fuel flow, operating pressure, and oxidizer temperature) for 100% power, it was possible to determine the percentage of the air flow that enters the combustion chamber through the swirlers and all the boundary conditions for 100% power.

In phase 2, also through an iterative process (simulation and result analysis), the available data presented by Ribeiro [28] regarding the operating conditions obtained in the test-bed charts presented (mainly the operating pressure and oxidizer temperature), the values of the ICAO's database [12], and the percentage of the air flow that enter the combustion chamber through the swirlers calculated in phase 1, it was possible to define the boundary conditions for 7% power. Finally, after definition the boundary conditions for the conditions of 7% and 100% power, the values of the operating pressure, oxidizer (air) temperature, overall AFR, and primary zone AFR for the other operating conditions were calculated through a linear regression. This allowed the obtention of the boundary conditions for the 30 and 85% power conditions.

Knowing the overall AFR and fuel flow for each condition, it was possible to determine the total air mass flow and then through the primary zone AFR, it is possible to calculate the air mass flow that enters the combustion chamber through the swirlers and the air mass flow that enters through the other entries (mixers and dilution holes). Once these steps were concluded, the boundary conditions for the air mass flow inlets were defined for each power condition of the LTO cycle.

5. Results and discussion

In this section, the results of the simulations are presented. It should be noted that all the results presented are only for one-fourth of the CFM56-3 combustor.

5.1 Combustor exit temperature

The first set of results was obtained during the process of estimation of the boundary conditions where the outlet average static temperatures to the simulations with Jet A fuel were calculated, as shown in **Figure 4**. Those are important values once we need them, first to compare with the reference temperature of 1649.94 K obtained by Ribeiro [28] to the condition of 100% power, and then to get reference values for the outlet temperature to allow the calculation of the mass flow of hydrogen fuel for each power condition.

The quantity of fuel for the simulations with hydrogen was calculated through the mass of Jet A fuel for each power condition and the ratio between the lower heating values, LHV, of the fuels, as shown in Eqs. (13) and (14).

$$\dot{m}_{Hydrogen\,fuel} = \dot{m}_{JetA\,fuel} \times \frac{LHV_{Jet-A}}{LHV_{Hydrogen}} \tag{13}$$

$$\dot{m}_{Hydrogen\,fuel} = \dot{m}_{JetA\,fuel} \times 0,3597 \tag{14}$$

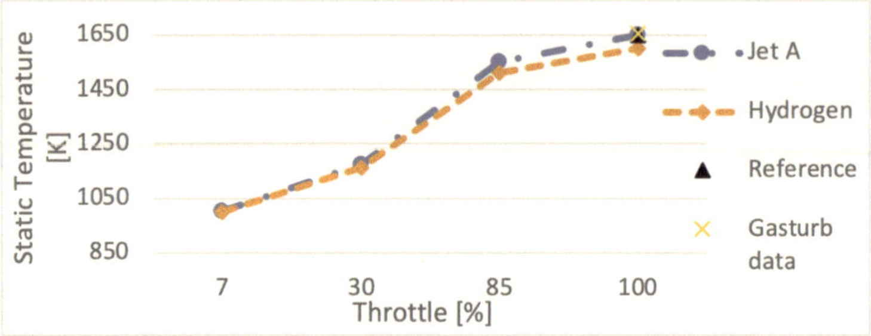

Figure 4.
Combustor outlet average static temperature throughout ICAO's LTO cycle, while burning Jet A and Hydrogen fuel; and the reference values for 100% throttle.

Since the LHV changes from author to author, this ratio was calculated and then adjusted for the condition of 7% power in order to get the same value for the outlet average static temperature for the simulations with the Jet A and hydrogen fuel.

Figure 4 shows the most relevant outlet average static temperatures obtained in this work. The values for the sensibility tests are not represented here.

Through **Figure 4** it is possible to conclude that the values of outlet temperature obtained through the simulations with the input data from ICAO and *GasTurb* values (blue line and yellow x, respectively) are quite similar to the reference value for the full power condition (to Jet A).

Through the same figure, but now considering the simulations with hydrogen, it is possible to conclude that:

- The values of outlet temperature for 7% and 30% power are similar to Jet A, which indicates that the ratios between the amount of hydrogen fuel and Jet A must be correct;

- For the conditions of 85% and 100% power, the lower outlet temperature for hydrogen fuel indicates that the quantity of hydrogen should be higher.

In this study, it was opted to fix only one value for the ratio, but as we can see, the best approach seems to be to calculate the specific ratio between the fuels to get the same exact outlet temperatures for each power condition.

5.2 NOx emissions

In the emission analysis, it is possible to study the emissions of all the pollutants, mainly CO, CO_2, UHC, and NOx. However, since in this work the only objective related to the pollutants is to compare the pollutant emissions between the Jet A fuel and hydrogen fuel, the only emissions analyzed were the NOx, once H_2O is not assumed as a pollutant.

Some of the results presented in this section were obtained through the emission index, EI. This value can be obtained using **Eq. (13)**:

$$EI[g/kg] = \frac{\text{Emission flow rate}[\text{kg/s}] \times 1000}{\text{Inlet } \dot{m}_{fuel} \ [\text{kg/s}]} \tag{15}$$

In **Eq. (13)**, the emission flow rate was obtained by reporting the *Flow rate* of the desired pollutant in the outlet plane in Ansys Fluent, and this value is given in kg/s. The values used for the $\dot{m}_{fuel}$ (obtained from ICAO [12]) at the inlets are given in kg/s. The results are presented in the form g[Emissions]/kg[fuel], which allows the comparison with the ICAO's reference data.

In the first analysis (reference standard), regarding Jet A fuel simulations, the emission index is used since the quantity of fuel burned is the same for each operating condition, and it can be easily interpreted by the reader. However, in the comparison between the pollutant emissions of Jet A fuel and hydrogen fuel, as the quantities of fuel are very different (the mass flow of hydrogen is almost a third of the mass flow of Jet A), the use of the EI may give a wrong perception of the emissions difference to the reader. So, to simplify the analysis, it was opted for the use of the flow rate of NOx, in grams per second, for a fourth of the chamber.

5.2.1 Control simulation and emission comparison between Jet A and hydrogen fuel

The results for the control simulations are presented in **Figure 5a**, which shows the EI(NOx) in the outlet of the quarter of the combustion chamber for the Jet A fuel.

As expected, and previously referred, for both approaches, the NOx emissions are lower at low power settings and attain maximum values at the highest power condition, where the temperatures are higher.

Considering **Figure 5a**, it is possible to conclude that:

- The NOx sub-mechanism clearly overpredicts the NOx quantity, in relation to the reference values (ICAO's database);

- Regarding to the NOx model available in ANSYS Fluent, for the lower power conditions (7 and 30% power), the model can predict values for the NOx

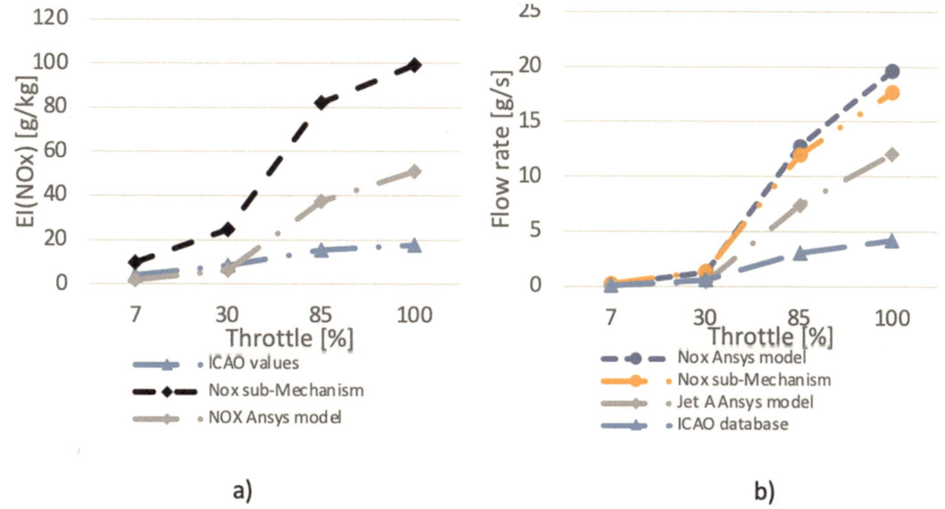

Figure 5.
Results for the NOx emissions for one-fourth of the combustion chamber throughout ICAO's LTO cycle: (a) EI results of NOx for the combustion of Jet A obtained through the use of two approaches, NOx sub-Mechanism and Ansys NOx model; and (b) comparison of the flow rates of NOx for the combustion of Jet A and Hydrogen fuel obtained through the use of two approaches, NOx sub-Mechanism and Ansys NOx model.

quantity close to the ones of ICAO's database; however, for the higher power conditions (85% and 100% power), this model also overpredicts the NOx emissions, in relation to the reference values (ICAO's database), approximately 2.4 times more for 85% power and 2.9 times more for 100% power.

- The NOx model from ANSYS Fluent can predict the NOx emissions better than the other approach, and these values will be used in the next step.

In a second phase, with the same boundary conditions (air mass flow rate, temperature, and pressure) for each operating condition, the simulations were repeated, but now for the hydrogen fuel. **Figure 5b** shows the comparison between the reference values (from ICAO and from the better model of the previous phase for Jet A fuel) and the NOx emission analysis made through two different approaches for hydrogen fuel, the NOx sub-mechanism, and the NOx model from ANSYS Fluent. As referred before, those analyses were made in terms flow rate, with the units of grams per second.

Looking at **Figure 5b**, considering the models used for NOx forecast, it is possible to take two principal conclusions:

- The forecasts made by the ANSYS NOx model provide higher values than those made with the NOx sub-mechanism for all the operating conditions; however, the error between these values is relatively small (up to 10%).

- Comparing the quantity of NOx produced by the hydrogen fuel with Jet A, for the lower power conditions, the quantities of NOx emitted are nearly the double of the values of the Jet A (for both ICAO's database and for the prediction with the model) and for the higher operating conditions, the quantity of NOx emitted continues to be nearly the double of the predicted NOx emissions for the Jet A fuel simulations with the NOx model and the emissions are predicted to be nearly four to five times higher than the reference data from ICAO.

5.2.2 Sensibility tests

In this work, several sensibility tests were carried out, namely the presence of the swirl effect on the swirler's inlets and the influence of the fuel injection pressure and temperature. In this work, the initial value of fuel temperature (used in the study) for the Jet A and the hydrogen fuel was 298.15 K, once this is the fuel temperature used by the software *Gasturb*. The results of these tests are presented in **Figure 6**.

As the represented values are similar, they are also presented in **Table 2**.

About the influence of the swirl effect, whose results for the NOx emissions are presented in **Figure 6a**, if one considers the same forecast approach (only the values of ANSYS model or the values of the sub-mechanism), it is possible to conclude that for this specific case the presence of this phenomenon helps to reduce slightly the quantity of NOx emissions for the high power conditions, while for lower power conditions the NOx values are closer.

To analyze the influence of the inlet fuel temperature, the sensibility tests were made with the hydrogen fuel at 600 K, and the results for the NOx emissions are presented in **Figure 6b**. From this figure, it is possible to conclude that:

- For the simulations with the hydrogen temperature of 600 K, the error between the two approaches used to forecast the NOx is considerable for the higher power

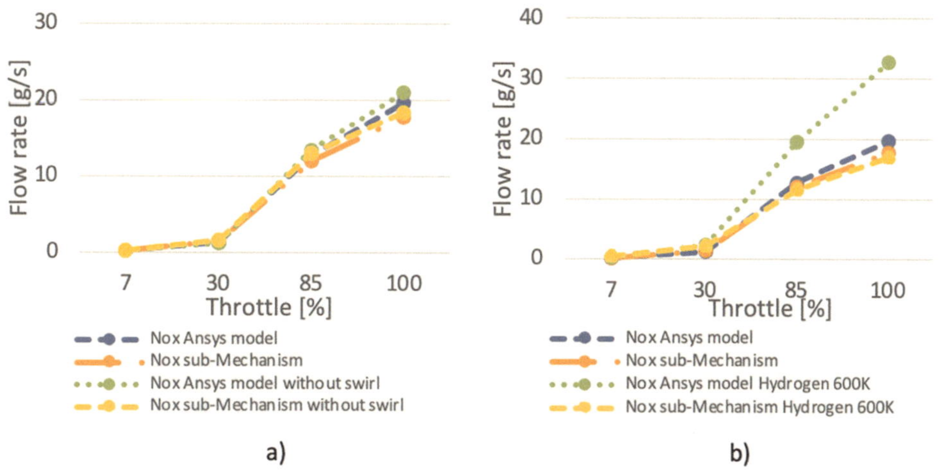

Figure 6.
Flow rates of NOx emissions for the sensibility tests for one fourth of the combustion chamber throughout ICAO's LTO cycle burning hydrogen: (a) analysis of the influence of swirl effect through the use of two approaches, NOx sub-mechanism and Ansys NOx model; and (b) analysis of the influence of the hydrogen fuel temperature through the use of two approaches, NOx sub-mechanism and Ansys NOx model.

Op. Cond.	Standard Reference		Without swirl		Fuel 600 K	
	Sub-Mechanism	***Ansys model***	***Sub-Mechanism***	***Ansys model***	***Sub-Mechanism***	***Ansys model***
100	17.70	19.58	18.30	20.91	16.97	32.70
85	11.97	12.75	12.98	13.40	11.51	19.49
30	1.44	1.30	1.61	1.22	2.26	2.38
7	0.22	0.17	0.24	0.16	0.51	0.32

Table 2.
*NOx flow rates in [g/s] obtained for hydrogen fuel (as in **Figure 6**).*

conditions (85 and 100% power); that is, if we look to the values obtained, the predicted emissions for the NOx model of Ansys are almost twice the value predicted with the NOx sub-mechanism.

- For both approaches, in the lower power conditions (7 and 30% power), the quantity of NOx emitted is higher for the fuel temperature of 600 K than for the reference temperature (298.15 K).

- For the higher power conditions (85 and 100% power), the approaches showed different behavior. For the sub-mechanism, the NOx emissions are lower for the fuel temperature of 600 K and for the NOx model of Ansys, the emissions keep higher than for the reference temperature (298.15 K).

About the injection pressure tests, the modification of this value did not make any changes to the results. For that reason, the results are not presented once they do not allow to analyze the influence of this parameter.

6. Conclusion

In this work, an overview of the use of hydrogen in aviation, the modifications needed to adapt an existent gas turbine to use hydrogen, and a CFD simulation of the CFM56-3 combustor burning hydrogen is provided. During this work, it was demonstrated theoretically that the CFM56-3 can work with hydrogen fuel with minor changes (related only to the injection system).

Regarding the results, starting with the control simulation (reference standard), there are several possible reasons that can be pointed out for the differences between the simulations and the ICAO's database values. For instance, the fact that the fuel was considered in the gaseous state when injected into the combustion chamber simulates "perfect" atomization, increasing the combustion efficiency and creating a higher temperature inside the combustion chamber. Other reason can be the fact that the chosen mechanism/sub-mechanism does not represent the combustion of the Jet A fuel or the NOx production in the best way. The choice of the radiation model can also influence this result, once the radiation representation is more important in the hydrocarbon fuel burn than in the hydrogen fuel burn. Other possible reason could be the chemical model used. However, due to the limited computational resources, it was not possible to use more complex models.

Regarding the other results, comparing the NOx emissions obtained for the simulations (for both Jet A and hydrogen), it was shown that for this geometry of combustor and injector, the quantity of NOx produced when burning hydrogen is almost twice of the NOx emissions for Jet A. Once we are using the same swirlers and injector geometry (single hole) for both fuels, these results are in agreement with the results of C. J. Marek et al. [30], who concluded in their job that using similar injection geometries, the minimum NOx levels for hydrogen fuel were twice than for Jet A fuel.

Finally, regarding the sensibility tests (changing the swirl effect, fuel injection pressure, and temperature), only the changes in swirl effect and the fuel temperature produced relevant changes in the results. The fact that the changes in fuel injection pressure did not produce major changes in the NOx emission results could be explained as the fact that for these tests, the geometry of the injectors was always considered the same (the area of the inlets did not change), and the fuel mass flow rate was the same for each power condition. For that reason, the pressure changes will not affect the behavior of the fuel jet that much. In practice, in gas turbine engines, the pressure is usually used to control the quantity of fuel injected.

Through the results obtained for the tests with the swirl effect, it was demonstrated that without the swirl effect, the NOx emissions increased. After analyzing the recirculation zone (position and form), it was concluded that without the swirl effect, the quality of the recirculation zone was reduced and the temperature across the combustion chamber was slightly increased (also increasing the emissions). So, the presence of the swirl effect helps to stabilize the recirculation zone, reducing the presence of hot spots in the flame.

About the influence of the fuel temperature, it was expected that an exponential raise in this value could affect largely the temperature in the outlet of the chamber, increasing the efficiency. However, that did not happen, and not only the outlet temperature changed just a small percentage (1–2%) for the double of the initial fuel temperature, but also cause malfunctions across the chamber, with greater evidence in the higher power conditions, where the analysis of the flame shows a great deterioration of the recirculation zone and the presence of a phenomenon that seems to be the occurrence of autoignition or flashback inside the swirlers. These malfunctions were associated with the velocity of hydrogen fuel flow, which has almost doubled when the temperature was raised. The most credible reason for this phenomenon consists of the density reduction of the hydrogen fuel due to that change in temperature.

Acknowledgements

This work is supported with Portuguese national funds by FCT - Foundation for Science and Technology, I.P., within the C-MAST - UIDB/00151/2020.

Author details

Rafael Domingues[1] and Francisco Brójo[2*]

1 C-MAST—Center for Mechanical and Aerospace Science and Technologies, Portugal

2 Department of Aeronautical Sciences, University of Beira, Covilhã, Portugal

*Address all correspondence to: brojo@ubi.pt

References

[1] ICAO (International Civil Aviation Organization). Global environmental trends – present and future aircraft noise and emissions, Working Paper, reference A40-WP/54 EX/21, 5/7/19

[2] ICAO (International Civil Organization). Aviation and the Environment: Outlook, Chapter one in ICAO 2019 Environmental Report, 2019

[3] European Commission. Revision of the EU Emission Trading System Directive 2003/87/EC concerning aviation [Internet], Ref. Ares (2020) 3515933, 03/07/2020. Available from: https://ec.europa.eu/clima/eu-action/transport-emissions/reducing-emissions-aviation_en [Accessed: March 17, 2022]

[4] ICAO (International Civil Organization). Sustainable aviation fuels guide, version 2. 2018

[5] ATAG (Air Transport Action Group). Waypoint 2050 report. Second Ed., September 2021

[6] Brewer GD. Hydrogen Aircraft Technology. Vol. 33431. Boca Raton, Florida: CRC Press; 1991

[7] Seeckt K. Conceptual Design and Investigation of Hydrogen-Fueled Regional Freighter Aircraft, [Licentiate Thesis]. Sweden: Stockholm; 2010

[8] Westenberger A. Hydrogen fuelled aircraft, LAirbus Deutschland GmbH 21129 Hamburg, Germany, 22/05/2003, IN: AIAA 2003-2880

[9] ENABLEH2 [Internet]. Cranfield University; 2022. Available from: https://www.enableh2.eu/ [Accessed: April 25, 2022]

[10] AIRBUS. ZEROe [Internet]. Available from: https://www.airbus.com/en/innovation/zero-emission/hydrogen/zeroe [Accessed: 2022-04-23]

[11] European Aviation Safety Agency. EASA TYPE-CERTIFICATE DATA SHEET -CFM56-2 & CFM56-3 Series Engines. TCDS: E.066, November, 2008

[12] ICAO. ICAO engine exhaust emissions data bank "CFM56-3-B1," Tech. Rep., January. 2015

[13] Aubuchon D, Campbell J. CFM56–3 Turbofan Engine Description. Toronto, Canada: Seneca College; 03/2016

[14] Oliveira J. CFD Analysis of the Combustion of Bio-Derived Fuels in the CFM56-3 Combustor [Master's Thesis]. Covilhã, Portugal: University of Beira Interior; 2016

[15] Mathur ML, Sharma RP. Gas turbines and jet and rocket propulsion. Second ed. Nem Chand Jain, Delhi; 2007

[16] El-Sayed AF. Aircraft Propulsion and Gas Turbine Engines. Second ed. Boca Raton, FL: CRC Press; 2017

[17] Lefebvre AH, Ballal DR. Gas Turbine Combustion. Third ed. Boca Raton, FL: CRC Press; 2010

[18] Morão I. The influence of jet fuels on the emission of pollutants [Master's Thesis]. Covilhã, Portugal: University of Beira Interior; 2019

[19] Lackner M, Winter F, Agarwal AK. Handbook of Combustion: Gaseous and Liquid Fuels. Vol. 3. KGaA, Weinheim: WILEY-VCH Verlag GmbH & Co; 2010. ISBN: 978-3-527-32449-1

[20] Haglind F, Singh R. Design of aero gas turbines using hydrogen. Journal of Engineering for Gas Turbines and Power. 2004;**128**:754-764

[21] Boggia S, Jackson A. Some unconventional aero gas turbines using hydrogen fuel. In: Proceedings of ASME Turbo Expo 2002, Amsterdam. Vol. 2B. New York: ASME; 2002. pp. 683-690

[22] Kuo KK. Principles of Combustion. Second ed. Hoboken, New Jersey: John Wiley & Sons, Inc.; 2005. ISBN 0-471-04689-2

[23] Mostafa, Mohammad Golam, 3D Simulation Of Jet-A Combustion In A Model Aircraft Engine Combustion Chamber. 2012. Theses 90

[24] Kundu K, Penko P, Yang S. Simplified Jet-A/air combustion mechanisms for calculation of Nox emissions. In: Proceedings of the 34th AIAA/ASME/SAE/ASEE Joint Propulsion Conference and Exhibit, Cleveland, Ohio, USA. 1998

[25] Ströhle J, Myhrvold T. An evaluation of detailed reaction mechanisms for hydrogen combustion under gas turbine conditions. International Journal of Hydrogen Energy. 2007;**32**:125-135. DOI: 10.1016/j.ijhydene.2006.04.005

[26] Briones AM, Aggarwal SK. A numerical study of H2–air partially premixed flames. International Journal of Hydrogen Energy. 2005;**30**:327

[27] ANSYS, Inc. 5 Best Practices for Gas Turbine Combustion Meshing Using Ansys Fluent. White Paper. 2020

[28] Ribeiro P. Análise de performance da Família de Motores de Avião CFM56. Master´s Thesis. Lisboa, Portugal: Instituto Superior de Engenharia de Lisboa; 2012

[29] ANSYS, Inc., ANSYS Fluent Theory Guide. Release 2020 R2, July 2020

[30] Marek CJ, Smith TD, Kundu K. Low emission hydrogen combustors for gas turbines using lean direct injection. In: 41st AIAA/ASME/SAE/ASEE Joint Propulsion Conference and Exhibit. Tucson, Arizona, USA. 2005

Chapter 5

Hydrogen Oxyfuel Combustion for Energy-Intensive Industries

Esin Iplik, Martin Adendorff and David Muren

Abstract

Hydrogen has been seen as a decarbonization enabler for the last few decades, and in the last couple of years, there have been many investments in its production through renewables and use in different industrial applications. It is often researched for energy storage, and combustion is an excellent alternative to recover the energy stored in hydrogen. It might be the most viable alternative, especially when it comes to energy-intensive metal and glass production processes. The utilization of hydrogen as a fuel in these processes would reduce greenhouse gas emissions significantly, considering their share in total emissions. Since these industries already benefit from oxyfuel combustion with traditional fuels for fuel savings, part of the infrastructure already exists for hydrogen oxyfuel combustion. Fuel change is expected to require some minor adjustments other than simply changing the oxidizer; however, each industry has specific points to consider. This chapter investigates metal and glass production processes based on their needs and challenges in using hydrogen oxyfuel combustion for heating. Additionally, possible exhaust gas stream improvements are suggested to recover energy and reduce emissions. Finally, safety aspects of hydrogen and oxygen use are discussed together with the community acceptance of hydrogen use.

Keywords: hydrogen, oxyfuel combustion, hydrogen combustion, decarbonization, energy-intensive

1. Introduction

Increasing renewable energy penetration in the electricity grid brought the discussion of power-to-x. Energy storage concepts are necessary for efficient energy utilization and also for grid stability, especially for uncontrollable and intermittent solar and wind energy [1]. Most of the power-to-x methods use electrolyzers as a core energy converter and suggest hydrogen usage within Haber–Bosch, methanation, Fischer–Tropsch, or other processes [2]. These strategies are classified as power-to-chemicals, power-to-liquid, power-to-heat, and power-to-power, based on their end use. At first, power-to-gas was a general term to describe the energy stored as a gas, excluding the end use. However, due to the different routes suggested, the use of power-to-x was expanded to include them all. Web of Science (www.webofknowledge.com) search for "power-to-x" gives 279 results, and over 50% of them are from

Germany. A search for "power-to-gas" results in 1736 scientific publications, and again Germany is the major contributor with 377 of the listed results. Finally, the "power-to-heat" search gives 471 results, and the yearly distribution of the articles with the given search terms is given in **Figure 1**. There is an increasing interest in heat demand and using stored renewable energy for heat. While part of the research is about district heating or residential heating, industrial heat demand is also investigated.

As of 2022, the European Clean Hydrogen Alliance has over 750 projects in its pipeline, 172 of these projects have an end use of hydrogen in industry. The main themes are the use of hydrogen for industrial heating, steel manufacturing, feedstock for ammonia, and synthetic fuels. Since these are application projects rather than research and are usually driven by industry, they show commitment, a level of maturity of relevant technologies, and investment possibilities (www.ech2a.eu).

Power-to-heat is an integral part of the discussion for industry because heat demand is roughly 70% of the total industrial energy demand in Europe [3]. While the wood industry needs 60% of the process heat around 80–100°C, ferrous and nonferrous metal processes require almost the same amount of their process heat over 1000°C [4]. Industrial heating with hydrogen can be achieved *via* fuel cells followed by electric heating or hydrogen combustion. Direct electrical heating should always be considered when it is technologically and economically feasible. However, since there are some technical challenges with high temperatures, heat generation by combustion is relevant for several industrial applications. While electrification and decarbonization of electricity production would decrease emissions significantly, hydrogen combustion can fill the gap in high-temperature heat demand. In 2019, around 38% of the heat demand in Europe was supplied by natural gas and other gasses in Europe, followed by 30% renewables and over 20% by solid fossil fuels [5].

The core of power-to-x and green hydrogen production relies on electrolyzers using renewable energy. There are three major types of water electrolyzers. The installed capacity and number of publications of each method are given in **Table 1**. The alkaline electrolyzer is the most mature and established technology. The installed capacity was 176 MW in 2020, and it still finds research ground. There were 576 articles published in 2020 about this technology; over 50% of the articles were from China. Proton exchange membrane (PEM) electrolyzers have an 89 MW installed capacity, and with its short

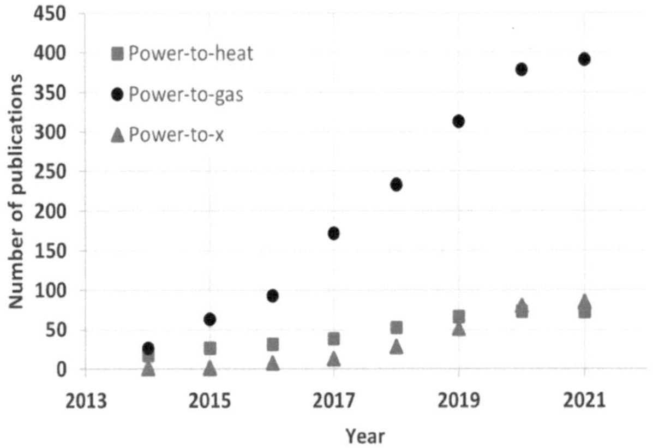

Figure 1.
Number of articles found on Web of Science for power-to-gas, power-to-x, and power-to-heat concepts.

Electrolysis method	Installed capacity (MW) [6]	Number of publications[a]
Alkaline electrolysis	176	576
PEM electrolysis	89	268
Solid-oxide electrolysis	1	142

[a]*Reconciled from [7].*

Table 1.
Major water electrolysis routes, their installed capacity, and the number of publications in 2020.

response time, PEM electrolysis is well suited to renewable energy storage. China and Germany were at the top of the list of countries contributing to the 268 articles on this technology. Solid-oxide electrolyzers are high-temperature systems and are considered to be advantageous for integration into high-temperature industries. With 142 publications, still high investment costs, short life cycles, and only around 1 MW of installed capacity, this technology is not considered to be commercially viable yet. Further details and technological advances can be found in recent review papers about alkaline [8], PEM [9], and solid-oxide [10] electrolyzers.

Oxygen is also produced with water electrolysis; however, the focus has been on the hydrogen side up to now, and oxygen is not captured. Industrial oxygen is produced from the air *via* air separation technologies that consume electricity and then used in wastewater treatment, combustion, and other applications. Oxygen-enriched combustion and oxyfuel combustion are not new concepts for some industries. Oxyfuel combustion is preferred for fuel saving or capacity increasing. Fuel saving up to 50% is possible depending on the flue gas temperature. Eliminating the nitrogen from the oxidizing mixture decreases the flue gas volume significantly, decreasing the flue gas's heat loss. While hydrogen is not a cheap fuel, its oxyfuel combustion is an excellent way to reduce costs.

An illustration for airfuel combustion and oxyfuel combustion for hydrogen is shown in **Figure 2**. While airfuel can melt roughly 3 ton/h aluminum burning 1200 Nm^3/h hydrogen with air, oxyfuel can achieve this production rate with 780 Nm^3/h hydrogen. It can also be used to increase capacity, and in that case, the fuel flow rate is kept constant at 1200 Nm^3/h to melt aluminum at 6 ton/h rate. Of course, converting from airfuel to oxyfuel requires investment, and there is also the continuous cost of oxygen, but usually, a new furnace installation costs more to increase the capacity.

This chapter explains hydrogen oxyfuel combustion, its chemistry, possible applications, and safe handling. For the applications, steel, aluminum, and glass industries are selected. The reason is the high-temperature requirements of the processes and their emissions. The traditional fuel choices of these processes vary between natural gas, different crude oil-based fuels, and LPG depending on the country and the facility. The steel industry has one of the highest greenhouse gas emissions, which accounts for around 7% of the global emissions [11]. Aluminum production accounts for roughly 1% of the global emissions, and aluminum has a higher energy consumption per kg product than steel [12]. Compared with metals, glass has a lower emission contribution [13]; however, considering the even higher process temperatures, a limited possibility for complete electrification (based on currently available technology), and to protect the optical properties of the product, a limited possibility to use solid biomass [14], hydrogen oxyfuel combustion could make a difference for this industry.

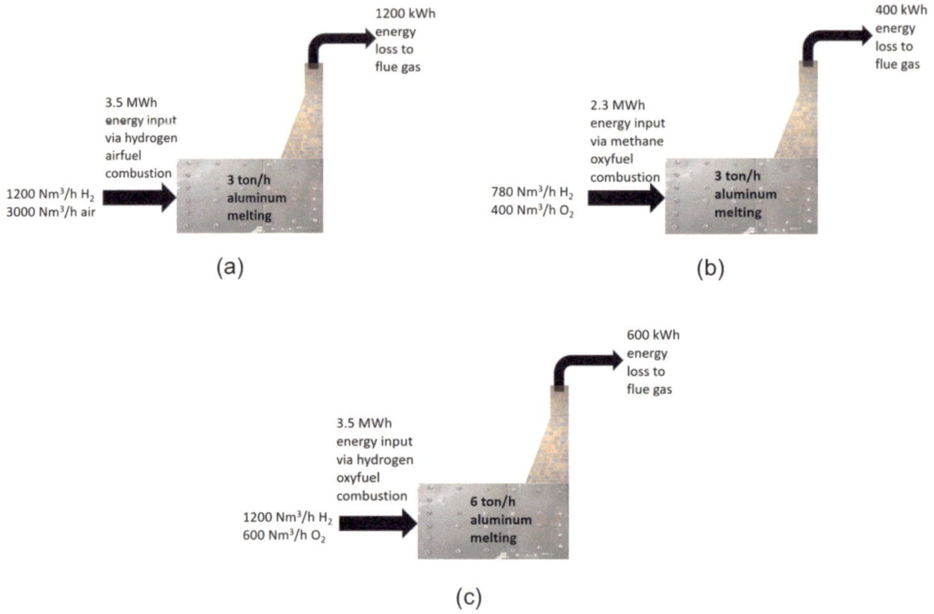

Figure 2.
Comparison of airfuel and oxyfuel combustion for aluminum melting. Note: These values are rough estimates and can vary between furnaces. No air preheating is considered. (a) Hydrogen airfuel combustion. (λ = 1.1), (b) Hydrogen oxyfuel combustion with lower fuel needs. (λ = 1.03), (c) Hydrogen oxyfuel combustion with higher capacity. (λ = 1.03).

2. Hydrogen oxyfuel combustion chemistry

Combustion of hydrogen with pure oxygen is a single reaction, given in Eq. (1), and the product is water.

$$H_2 + \frac{1}{2}O_2 \rightarrow H_2O \quad (1)$$

Hydrogen oxyfuel combustion has other reactions if the furnace is not air tight. While in a laboratory, tightness is highly likely to be achieved, under practically all industrial conditions, air in-leakage into a furnace is the norm, and only the amount and location of the in-leakage can be controlled, based on the furnace pressure and process parameters. Energy efficiency and combustion efficiency are of course directly affected by these leaks.

Hydrogen combustion has the adiabatic flame temperature of 2377 K with air, and 3072 K with oxygen under stoichiometric conditions and atmospheric pressure [15]. The enthalpy of combustion is −241.8 kJ/mol H_2 [16]. If the furnace is air tight, without nitrogen or carbon present in the gas mixture, NOx and CO_2 emissions are of course zero. When the furnace leaks, undesired side reactions generate NOx, which is produced only *via* the thermal route (Zeldovich mechanism, given in Eq. (2), (3), and (4)) since the fuel does not contain nitrogen compounds. These undesired reactions can be significantly reduced by lowering the peak flame temperature, as a high temperature is required for oxygen and nitrogen dissociation reactions. Additionally, lower nitrogen partial pressure (less air drawn into the furnace) and lower oxygen partial pressure (no excess oxygen) limit these reactions. Further information on Zeldovich reactions can be found in [17].

$$N_2 + O \rightarrow NO + N \quad (2)$$

$$N + O_2 \rightarrow NO + O \quad (3)$$

$$N + OH \rightarrow NO + H \quad (4)$$

Industrial furnaces are typically operated with a slight positive pressure (in the range of a few mm of water column) to minimize air ingress. In this case, the amount of nitrogen in the furnace can be reduced to a great extent, and NOx emissions can be minimized. Specific furnace designs as well as frequent charging operations on some types of furnaces and the associated air ingress, however, can make NOx control very challenging. The problems associated with these leaks are not limited to the emissions, as the combustion efficiency and furnace temperature also suffer, requiring increased energy input to maintain operating conditions, while the product quality may suffer as well.

Along with oxyfuel combustion, O_2-enhanced combustion (sometimes referred to as oxygen enrichment) is sometimes discussed in the literature [18]. Besides the potential for increased NOx generation, it brings with it certain operational drawbacks, especially if the enrichment level and power load changes are required during operation. These processes must be sized for airfuel combustion, which has the highest flue gas flow rate. As the O_2 enrichment level increases, the flue gas flow rate decreases significantly, and this fact limits the lowest possible power level, while still maintaining a suitable furnace pressure. Complete oxyfuel combustion offers a simpler control of the process conditions if the flue gas system is appropriately redesigned.

Maintaining the furnace pressure is vital to prevent leakage, and it is directly dependent on the flue gas flow rate. The difference between natural gas oxyfuel and hydrogen oxyfuel in terms of flue gas volumes is insignificant when compared with the difference between airfuel and oxyfuel. If the flue gas duct is sized based upon natural gas oxyfuel combustion, maintaining the furnace pressure of hydrogen oxyfuel operation would be no concern. Of course, there are some process-specific requirements, and the following two sections discuss two metal processes and glass production for hydrogen oxyfuel combustion.

3. Metal production processes

3.1 Steel production

Steel production is based on ore (primary) or scrap (secondary) processing. The primary production starts with a blast furnace producing pig iron, followed by a basic oxygen furnace that produces steel by oxidizing the excess carbon and other undesired elements. A blast furnace uses coke for the reduction of iron oxides. There are also direct reduction applications with natural gas and hydrogen, but the blast furnace route dominates the primary steel production. Secondary production is carried out in electric arc furnaces (EAFs) that are charged with scrap material. Both of the routes have high-temperature requirements.

The iron and steel industry has higher greenhouse gas emissions than other industrial processes. China has the highest steel production from ore, which is roughly 66% of the world's total. EU-28 countries, Japan, and South Korea follow China in primary production [19]. Secondary production is a quarter of total steel production [20], and the leading region is North America, followed by EU-28 countries and India. Both routes can utilize hydrogen oxyfuel combustion in auxiliary and downstream processes, that is, for ladle preheating and reheating furnaces. In fact, secondary production already uses natural gas oxyfuel combustion. However, the world's total hydrogen production capacity is not enough to decarbonize the entire industry. In total, only 4% of the global hydrogen production relies on water electrolysis [21] and the rest is produced through fossil fuel-based routes, such as steam methane reforming. All leading steel-producing countries have green hydrogen policies and investment plans; however, usually, these plans include the transport and mobility sector instead of the industrial applications. Noteworthy green hydrogen integration in steel production projects takes place in Sweden and Germany, namely, Hybrit [22], H2 Green Steel [23], and Salcos [24].

For both primary and secondary productions, oxyfuel combustion to supply high-temperature heat demand is discussed in the literature. Although the secondary production is based on EAFs, oxyfuel burners are frequently included to shorten the cycle time and to reduce electricity consumption [25]. For primary steel production, there is blast furnace retrofitting suggestions in the literature with hydrogen injection [26] and coke oven gas recirculation (contains over 50% H_2 and 30% methane) [27], next to a more fundamental direct reduced iron route [28]. While the first one suggests using lower amounts of carbon compounds and instead includes hydrogen in the mix, the latter eliminates these substances and suggests using only the hydrogen itself for reduction. The reduction reactions of CO and H_2 are given below, and the H_2 reactions are faster.

$$3Fe_2O_3 + H_2 \rightarrow 2Fe_3O_4 + H_2O \quad (5)$$

$$Fe_3O_4 + H_2 \rightarrow 3FeO + H_2O \quad (6)$$

$$FeO + H_2 \rightarrow Fe + H_2O \quad (7)$$

$$3Fe_2O_3 + CO \rightarrow 2Fe_3O_4 + CO_2 \quad (8)$$

$$Fe_3O_4 + CO \rightarrow 3FeO + CO_2 \quad (9)$$

$$FeO + CO \rightarrow Fe + CO_2 \quad (10)$$

The investment impact would be higher in the primary production plants, considering their significantly higher capacity and much higher emissions. The alternative blast furnace design is expected to decrease the emissions as lower amounts of carbon compounds would be required, and the DRI route eliminates the emissions.

An important point to consider is product quality as changing the process gases, and the resultant different flue gas mixture might cause some undesirable side reactions. For steel production, hydrogen embrittlement is a nonnegligible concern when atomic hydrogen is present. If, however, the burner design and the furnace control are well performed, unburned hydrogen would not exist as a precursor for this problem. Another concern is the scale layer and its morphology. In hot rolling mills, a scale layer is formed on the steel, and then, it is removed by a descaler. If the morphology of the scale changes due to a change in the production, the descaler might cause defects on the surface. Tests at different temperatures and atmospheres show a slight difference in the scale thickness between natural gas and hydrogen combustion [29]. The same work shows no decarburization in hydrogen oxyfuel combustion conditions, which is promising. This, on the other hand, needs further investigation for high carbon steel as the information does not specify the samples' content. A full-scale trial run with hydrogen oxyfuel combustion for steel reheating showed no change in the scale formation next to other quality indicators [30].

3.2 Aluminum production

Primary aluminum production starts with bauxite, and alumina is extracted by the Bayer process and purified by the Hall–Heroult process afterward. The very high energy requirement of the primary production route further increases the importance of secondary production, which requires around 94% less energy per kg of aluminum [31]. In 2020, worldwide, aluminum had an over 30% recycle rate, which increases every year [32]. However, high-quality aluminum production requires more energy if produced *via* the secondary route [33]. Both routes have high-temperature requirements. China is the leading country of primary aluminum production with over 50% [34]. India, Russia, and North America are other important producers [35]. While China and Russia mainly rely on primary production, North America has a higher secondary production capacity [36]. In Europe, Germany, the United Kingdom, and Italy are the important secondary producers [37].

Oxyfuel combustion is used in aluminum melting and remelting processes and can reduce dross formation in primary and secondary production processes. There are very specific concerns related to hydrogen oxyfuel combustion due to the change in the furnace atmosphere. H_2 is very soluble in molten aluminum, and it might cause pinholes, passages, and blisters if remains dissolved when the molten aluminum solidifies [31]. This, of course, can be avoided by burner design and proper process operation. It should also be considered that the only source of H_2 dissolution is not the fuel. A water dissociation reaction occurs at the liquid aluminum surface and generates atomic hydrogen [38] and alumina, according to Eq. (11).

$$2Al + 3H_2O \rightarrow Al_2O_3 + 6H+ \tag{11}$$

Water is of course present at a very high partial pressure in the combustion gases of hydrogen oxyfuel combustion; therefore, further research and trial runs are essential to determine the effect and the importance of this reaction. However, dissolved H_2 is efficiently removed by the degassing process before casting [39]. Another concern is increased dross formation due to the previously given water dissociation and higher oxygen partial pressure in the furnace atmosphere, causing the reaction given in Eq. (12).

$$4Al + 3O_2 \rightarrow 2Al_2O_3 \tag{12}$$

CO_2 is known to decrease dross losses, and combined trials of hydrogen oxyfuel combustion and inert gas injection can be used to compare against a reference natural gas combustion scenario. Hydrogen as a reduction agent was also suggested to recover iron and aluminum from the by-product of the Bayer process, bauxite residue, and it was found successful in lab trials [40]. It is important to note that for emission-free aluminum production, the use of hydrogen for heating and decarbonization of the electricity used for the electrolysis should be complemented by carbon-free anodes instead of graphite for Hall–Heroult process since this electrolytic method contributes to the emissions significantly.

4. Glass production process

Glass production is energy-intensive in the melting and fining steps, with furnace temperatures of up to 1600°C. The melting process is primarily carried out in regenerative furnaces with a very small number using recuperative furnaces; both rely on heat recovery since there is significant waste heat in the hot flue gases. In some geographies, these furnaces have been replaced by full oxyfuel furnaces, usually either due to very strict environmental legislation or very high fuel prices. For specialty glasses, oxyfuel furnaces are not uncommon, and for small capacities, electric furnaces can be used; an electric furnace, however, has a shorter lifetime [13]. The glass production is mainly local since product transportation poses a high risk of loss. Recycling rates are high, especially for container glass in developed countries (78% in EU28 and 32% world). Unlike the metal industry, recycling and primary production are carried out in the same furnace. Emissions are different based on the production location, and currently, in Europe and North America natural gas is the fuel of choice. Over 60% of glass production emissions are due to combustion [41]. The rest of the CO_2 is formed in calcination reactions of soda ash and limestone given in Eq. (13) and (14).

$$Na_2CO_3 \rightarrow Na_2O + CO_2 \tag{13}$$

$$CaCO_3 \rightarrow CaO + CO_2 \tag{14}$$

A recent review on the decarbonization of the glass industry focuses on higher recycling rates and changing the fuel [42]. Due to the above reactions, glass production would still emit CO_2 with hydrogen oxyfuel combustion. However, the flue gas is expected to be more suitable for carbon capture methods with fewer soot particles and higher CO_2 concentrations after condensation of the water content.

There are particular product and furnace design concerns resulting from hydrogen oxyfuel combustion in glass melting. Foaming and, therefore, limited heat transfer is a primary concern. This also happens in existing operations and is solved by decreasing the air ratio to form CO that breaks down the foam. The high water vapor concentration of the furnace atmosphere is believed to exacerbate the issue. Another concern is a reduced furnace lifetime due to refractory damage caused or accelerated by the higher water vapor content.

There are a number of projects focusing on the use of hydrogen oxyfuel combustion for the glass industry investigating if these issues will occur and how they can be mitigated. There are projects in Germany and the U.K. testing hydrogen and enriched natural gas for glass melting that are HyGlass [43], Kopernikus [44], and HyNet [45]. By the end of 2021, HyGlass was finalized, and the trials had demonstrated the success of hydrogen airfuel combustion. The heat distribution was found homogenous with hydrogen combustion [46]. From the economic point of view, hydrogen is an expensive fuel for airfuel combustion. On an industrial scale, hydrogen oxyfuel combustion is expected to play a significant role, most likely in combination with electrical heating in the so-called hybrid furnaces, and the glass industry requires dedicated projects to ensure compatibility.

5. Economizer and latent heat utilization

Heat recovery is a crucial part of the aforementioned industries. Due to the high temperatures required for the process chemistry, the flue gases also have high temperatures. Current practices usually show partial waste heat recovery for basic oxygen, electric arc, and glass furnaces. Annual energy loss to flue gas in high-temperature applications is relatively high. Blast furnaces and electric arc furnaces have over 10% heat loss with the flue gas; this value is over 25% for glass melting furnaces and finally over 65% for aluminum melting furnaces [47]. Decreasing the amount of flue gas with hydrogen oxyfuel combustion will partly solve this problem, and the rest of the energy can be recovered using different systems. Waste heat recovery systems have certain restrictions due to materials, capital costs, and maintenance issues.

Heat pipes can be used for over 1000°C flue gas temperatures [48]. These systems utilize phase transition and are often used in solar collectors to keep the temperature difference low [49]. Tests in a highly corrosive environment show that with a suitable material selection, a long lifetime can be expected, and maintenance needs can be kept minimal [50].

Below 500°C, condensing economizers can recover the heat. As the name suggests, latent heat of the flue gas stream is also utilized in a condensing economizer. Usually installed by the district heat producers to increase the total efficiency [51, 52], condensing economizers are also discussed for the flue gas heat recovery. However, the heat recovered would be useful only for low-temperature heat demand. The risk of corrosion and deposition has research focused on coatings [53, 54]. One advantage of hydrogen oxyfuel combustion here is the complete elimination of sulfur gases, which reduces the risk of corrosion since sulfidation is one of the main corrosion mechanisms.

Additionally, the high water vapor content of the flue gas would provide a higher latent heat recovery. The lower volume and the higher water vapor ratio of the flue gas raise questions about condensation in the flue gas duct, but this can be used to cut energy losses with careful planning. A cleaner flue gas stream makes it easier to plan heat recovery systems and their maintenance.

6. Safety aspects and community engagement

Like all combustible gases, hydrogen has several important safety requirements. Due to a few prominent accidents in the past (Hindenburg and Challenger), it has a reputation for being very dangerous. Years after these accidents, thorough investigations have determined that hydrogen was not the reason behind them. When compared with many other conventional fuels, hydrogen has its benefits. Unlike gasoline or diesel, hydrogen is nontoxic and neither does it contaminate the environment. A leak in a confined space would decrease the O_2 concentration, resulting in an asphyxiation hazard. As it is lighter than air an outdoor leak is highly unlikely to create an explosive mixture; instead, it will disperse upward and escape in the atmosphere or, if ignited, it would burn until the hydrogen source has been removed. A hydrogen flame has low radiation due to its lack of carbon-containing combustion products, for example, soot. This fact decreases the risk of secondary fires but also makes it harder to detect a hydrogen flame as it is almost invisible when burning in the air in bright outdoor conditions [55], which necessitates fast detection of a leak in a confined space. The difference between a natural gas and hydrogen oxyfuel flame can be seen in **Figure 3**.

This broad flammability range makes a hydrogen flame very stable, and it is difficult to provoke a hydrogen flame to lift off of a burner. Its very low ignition energy also poses a risk, while at the same time simplifying the ignition of a hydrogen-fired burner. Anti-static clothes and tools should always be used when working with hydrogen or hydrogen-containing piping or equipment. In addition, all equipment

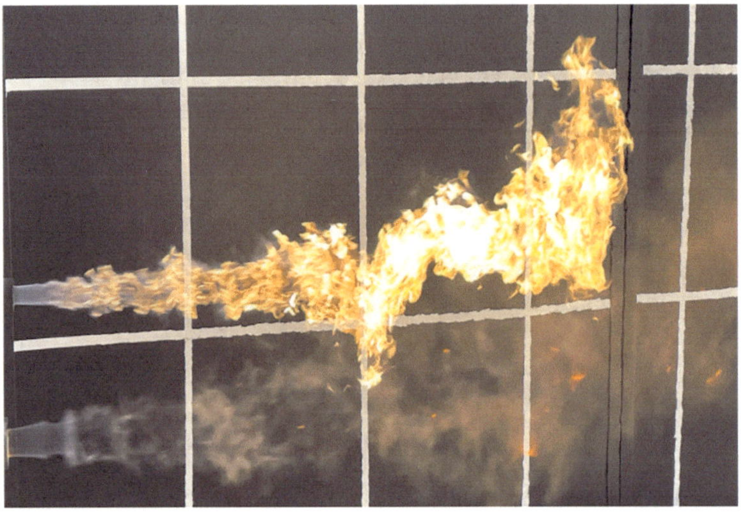

Figure 3.
Natural gas (upper) and hydrogen (lower) oxyfuel flames with the same burner power.

must be correctly grounded, and all vents and chimneys protected by a lightning conductor. As mentioned previously, hydrogen is a very light gas; therefore, adequate ventilation would decrease the risks. If adequate ventilation is always present, then small-to-moderate hydrogen leaks cannot accumulate. Additionally, hydrogen requires a high oxygen concentration to create a detonative mixture (18–59% [56]); therefore, even in a confined space, it is more likely to burn than to detonate.

Due to its very small molecular size, it can diffuse rapidly through some porous materials or systems that are considered gas tight with other gases. As a result, threaded connections and fittings should be avoided, or kept to an absolute minimum, while welded joints should be prioritized. Valves and flanges are susceptible to leaks and therefore require specialized sealing elements, for example, tongue and groove flanges. Ensure that all gaskets, seals, and lubricants are hydrogen compatible as hydrocarbons are cracked over time and purged out of the material. Some materials, such as cast iron, cannot be used for hydrogen piping elements since it loses ductility over time with diffusing hydrogen [57]. Copper and copper /tin- /zinc-based alloys should not be used for hydrogen service.

Although hydrogen is being discussed more often than ever due to environmental concerns, industrial applications have used hydrogen for many decades. Ammonia production is an excellent example of high-volume hydrogen usage showing that it is possible to use this gas without any problems if the appropriate safety requirements are complied with.

The safe handling of oxygen is frequently overlooked since it occurs naturally, and it is essential for our survival. High purity oxygen has very strict safety requirements, and accidents can result in very serious injuries and deaths. While oxygen is not flammable, many compounds that are practically inert in air, burn readily in oxygen. Even stainless steel can burn at high oxygen concentrations and pressures. The flammability range of hydrogen in oxygen is from 4–94% [58]. Breathing pure oxygen can cause dizziness, vision loss, and loss of consciousness. Oxygen saturation of clothing is a severe fire hazard since almost all clothing materials ignite readily and burn vigorously when saturated. All piping, components, and instruments in an oxygen system must be free of all hydrocarbons (e.g., oil and grease, to name some), organic contamination, dirt, moisture, and particles. Valves and rotating instruments have special requirements to eliminate potential ignition sources. Valves should always be opened very slowly to prevent ignition due to particle impingement or adiabatic compression. Magnesium and titanium alloys should be avoided for the oxygen-containing components and piping. Oxygen is slightly heavier than air and can therefore accumulate in pits, cellars, and underground rooms.

Since both hydrogen and oxygen are stored under pressure, the appropriate personal protective equipment (PPE) should always be used when working with them. Flame retardant and anti-static clothing, safety glasses, safety shoes, and portable gas analyzers are typically the minimum protective equipment used in industrial applications where hydrogen and/or oxygen are present. Atmospheric monitoring equipment should be positioned according to gas specifications and possible leak locations and should be calibrated and maintained regularly. If cylinders are used for the gas supply, these should be clearly marked, well secured, and correctly stored.

It is worth noting that using these gases together does not pose a significantly higher risk than other oxidizer fuel combinations. Hydrogen and oxygen do have a much broader combustion range and a mixture will ignite very easily, hydrogen has other features that make it “less dangerous” than other fuels; for example, LPG being significantly heavier than air will readily accumulate in underground rooms, cellars, sewers, and pits. Each gas has its own specific requirements for storage, transportation, and use and these must be fully complied with at all times. These requirements might differ, which does not mean that one is more dangerous than the other. Although most of it is not green, hydrogen is produced on an industrial scale, transported by trailers, rail, or pipelines, and has been safely used in many different industrial applications for decades. Oxygen is produced by air separation units and used in industry, hospitals, and homes for life support. Knowing and understanding the risks is the first step toward safety, and this should not cause anxiety but instead should bring confidence.

7. Conclusions

In the last decades, research efforts focused on hydrogen as an environmentally friendly energy carrier, which brought with it the discussion of its conversion back to energy. While hydrogen is used in fuel cells to generate electricity, its combustion provides thermal energy. Since green hydrogen is not a cheap fuel, hydrogen oxyfuel combustion can be used to reduce fuel consumption and the energy loss to the flue gases. The potential of three high-temperature industries is discussed for hydrogen oxyfuel combustion applications. The change in the furnace atmosphere and the possible product interactions are presented based on the literature. In this aspect, only a limited number of lab-scale experiments are documented for the aluminum and glass industries. Pilot-scale trials have the potential to show the full effect on products, energy gain, and environmental advancements. The iron and steel industry is demonstrating a dedicated effort toward carbon-free production methods. Certainly, their experience in handling hydrogen on a large scale will have an effect on the other industries. Thanks to the change in flue gas stream, possible gains are suggested to decrease emissions and increase heat recovery. These, however, require conceptual design and feasibility analysis. Finally, safety precautions for hydrogen and oxygen use are stated to run hydrogen oxyfuel combustion with the lowest risk possible.

Acknowledgements

This project has received funding from the European Union's Horizon 2020 research and innovation programme under grant agreement No 958307 (https://h2020harare.eu/).

Author details

Esin Iplik[1,2]*, Martin Adendorff[1] and David Muren[3]

1 Metals, Combustion and Energy, Linde Technology, Unterschleißheim, Germany

2 Department of Sustainable Energy Systems, Mälardalen University, Västerås, Sweden

3 Metals, Combustion and Energy, Linde Technology, Stockholm, Sweden

*Address all correspondence to: esin.iplik@linde.com

References

[1] Bailera M, Lisbona P, Romeo LM, Espatolero S. Power to gas projects review: Lab, pilot and demo plants for storing renewable energy and CO_2. Renewable and Sustainable Energy Reviews. 2017;**69**:292-312

[2] Wulf C, Linßen J, Zapp P. Review of power-to-gas projects in Europe. Energy Procedia. 2018;**155**:367-378

[3] Papapetrou M, Kosmadakis G, Cipollina A, La Commare U, Micale G. Industrial waste heat: Estimation of the technically available resource in the EU per industrial sector, temperature level and country. Applied Thermal Engineering. 2018;**138**:207-216

[4] Kosmadakis G. Estimating the potential of industrial (high-temperature) heat pumps for exploiting waste heat in EU industries. Applied Thermal Engineering. 2019;**156**:287-298

[5] Eurostat. Electricity and Heat Statistics. 2019. Available from: https://ec.europa.eu/eurostat/statistics-explained/index.php?oldid=401097msclkid=9b32bb5ea5ef11ec93d526e6885dcfa9#Derived_heat_production [Accessed: May 9, 2022]

[6] International Energy Agency. Global Hydrogen Review. 2021. Available from: https://www.iea.org/data-and-statistics/charts/global-installed-electrolysis-capacity-by-technology-2015-2020 [Accessed: May 9, 2022]

[7] Fuel Cells and Hydrogen Observatory. Available from: https://fchobservatory.eu/observatory/publications [Accessed: May 9, 2022]

[8] Brauns J, Turek T. Alkaline water electrolysis powered by renewable energy: A review. PRO. 2020;**8**(2):248

[9] Kumar SS, Himabindu V. Hydrogen production by PEM water electrolysis–A review. Materials Science for Energy Technologies. 2019;**2**(3):442-454

[10] Hauch A, Küngas R, Blennow P, Hansen AB, Hansen JB, Mathiesen BV, et al. Recent advances in solid oxide cell technology for electrolysis. Science. 2020;**370**(6513):eaba6118

[11] An R, Yu B, Li R, Wei YM. Potential of energy savings and CO_2 emission reduction in China's iron and steel industry. Applied Energy. 2018;**226**:862-880

[12] Gautam M, Pandey B, Agrawal M. Carbon footprint of aluminum production: Emissions and mitigation. Environmental Carbon Footprints. 2018:197-228. [Chapter ID: 8]

[13] Zier M, Stenzel P, Kotzur L, Stolten D. A review of decarbonization options for the glass industry. Energy Conversion and Management. 2021;**10**:100083

[14] Dreizler A, Pitsch H, Scherer V, Schulz C, Janicka J. The role of combustion science and technology in low and zero impact energy transformation processes. Applications in Energy and Combustion Science. 2021;7:100040

[15] Glassman I, Yetter RA, Glumac NG. Combustion. 5th Ed. Elsevier, Academic press; 2014

[16] Mills RL, Ray P, Dhandapani B, Nansteel M, Chen X, He J. New power source from fractional quantum energy levels of atomic hydrogen that surpasses internal combustion. Journal of Molecular Structure. 2002;**643**(1–3):43-54

[17] Tsuji H, Gupta AK, Hasegawa T, Katsuki M, Kishimoto K, Morita M. High Temperature Air Combustion: From Energy Conservation to Pollution Reduction. Taylor & Francis, CRC Press; 2002

[18] Dzurňák R, Varga A, Jablonský G, Variny M, Pástor M, Lukáč L. Analyzing the formation of gaseous emissions during Aluminum melting process with utilization of oxygen-enhanced combustion. Meta. 2021;**11**(2):242

[19] Wang P, Zhao S, Dai T, Peng K, Zhang Q, Li J, et al. Regional disparities in steel production and restrictions to progress on global decarbonization: A cross-national analysis. Renewable and Sustainable Energy Reviews. 2022;**161**: 112367

[20] Fan Z, Friedmann SJ. Low-carbon production of iron and steel: Technology options, economic assessment, and policy. Joule. 2021;**5**(4):829-862

[21] Zhang C, Greenblatt JB, Wei M, Eichman J, Saxena S, Muratori M, et al. Flexible grid-based electrolysis hydrogen production for fuel cell vehicles reduces costs and greenhouse gas emissions. Applied Energy. 2020;**278**:115651

[22] Hybrit. A Fossil-Free Future. 2016. Available from: https://www.hybritdevelopment.se/en/a-fossil-free-future/ [Accessed: May 9, 2022]

[23] H2 Green Steel. 2020. Available from: https://www.h2greensteel.com/ [Accessed: May 9, 2022]

[24] Salcos. Klimainitiative zur CO_2-armen Stahlproduktion. 2015. Available from: https://salcos.salzgitter-ag.com/ [Accessed: May 11, 2022]

[25] Von Schéele J. Oxyfuel combustion in the steel industry: Energy efficiency and decrease of CO_2 emissions. Energy Efficiency. 2010;**83**:83-102

[26] Yu X, Hu Z, Shen Y. Modeling of hydrogen shaft injection in ironmaking blast furnaces. Fuel. 2021;**302**:121092

[27] Higuchi K, Matsuzaki S, Saito K, Nomura S. Improvement in reduction behavior of sintered ores in a blast furnace through injection of reformed coke oven gas. ISIJ International. 2020; **60**:2218-2227

[28] Vogl V, Åhman M, Nilsson LJ. Assessment of hydrogen direct reduction for fossil-free steelmaking. Journal of Cleaner Production. 2018;**203**:736-745

[29] Cirilli F, Jochler G, Mosconi M, Praolini F. Effects of H2 combustion on scale growth and steel surface quality in reheating furnaces. Matériaux & Techniques. 2021;**109**(3–4):302

[30] Von Schéele J. Embracing hydrogen flameless oxyfuel for CO_2 free reheating. Iron & Steel Today. 2020;**10/11**:18-20

[31] Campbell FC. Metals fabrication: Understanding the basics. ASM International. 2013

[32] IEA. Aluminum Tracking Report. Available from: https://www.iea.org/reports/aluminium [Accessed: May 11, 2022]

[33] Padamata SK, Yasinskiy A, Polyakov P. A review of secondary aluminum production and its byproducts. Journal of Metals. 2021; **73**(9):2603-2614

[34] Yu B, Zhao Z, Zhang S, An R, Chen J, Li R, et al. Technological development pathway for a low-carbon primary aluminum industry in China. Technological Forecasting and Social Change. 2021;**173**:121052

[35] Primary Aluminum Production. Available from: https://international-aluminium.org/statistics/primary-aluminium-production/ [Accessed: May 11, 2022]

[36] National Minerals Information Center. North America, Central America, and the Caribbean. Available from: https://www.usgs.gov/centers/national-minerals-information-center/latin-america-and-canada-north-america-central-america [Accessed: May 11, 2022]

[37] National Minerals Information Center. Europe and Central Eurasia. https://www.usgs.gov/centers/national-minerals-information-center/europe-and-central-eurasia [Accessed: May 11, 2022]

[38] Xu H, Jian X, Meek TT, Han Q. Degassing of molten aluminum A356 alloy using ultrasonic vibration. Materials Letters. 2004;**58**(29):3669-3673

[39] Ren Y, Ma W, Wei K, Yu W, Dai Y, Morita K. Degassing of aluminum alloys via the electromagnetic directional solidification. Vacuum. 2014;**109**:82-85

[40] Pilla G, Kapelari SV, Hertel T, Blanpain B, Pontikes Y. Hydrogen reduction of bauxite residue and selective metal recovery. Materials Today: Proceedings. 2022;**57**:705-710

[41] PROTECTION E. Mandatory Reporting of Greenhouse Gases. RIN. 2009;**2060**:A079

[42] Pisciotta M, Pilorgé H, Feldmann J, Jacobson R, Davids J, Swett S, et al. Current state of industrial heating and opportunities for decarbonization. Progress in Energy and Combustion Science. 2022;**100982**

[43] HyGlass. https://www.bvglas.de/en/detail/news/bv-glas-and-gwi-complete-their-hyglass-project [Accessed: May 11, 2022]

[44] Kopernikus. Available from: https://www.kopernikus-projekte.de/ [Accessed: May 11, 2022]

[45] HyNet North West. Available from: https://hynet.co.uk/ [Accessed: May 11, 2022]

[46] Islami B, Giese A, Biebl M, Fleischmann B, Overath J, Nelles C. Wasserstoffnutzung in der Glasindustrie als Möglichkeit zur Reduzierung von CO2-Emissionen un des Einsatzes erneuerbarer Gase - Untersuchung der Auswirkungen auf den Glasherstellungprozess und Analyse der Potenziale in NRW. In: Ministerium für Wirtschaft, Innovation, Digitalizierung und Energie des Landes Nordrhein-Westfalen. 2022. Available from: https://www.bvglas.de/media/BV_Glas/HyGlass_Abschlussbericht_final.pdf

[47] Vance D, Nimbalkar S, Thekdi A, Armstrong K, Wenning T, Cresko J, et al. Estimation of and barriers to waste heat recovery from harsh environments in industrial processes. Journal of Cleaner Production. 2019;**222**:539-549

[48] Agathokleous R, Bianchi G, Panayiotou G, Aresti L, Argyrou MC, Georgiou GS, et al. Waste heat recovery in the EU industry and proposed new technologies. Energy Procedia. 2019;**161**: 489-496

[49] Bienert WB. The heat pipe and its application to solar receivers. Electric Power Systems Research. 1980;**3**(1–2): 111-123

[50] Hansen G, Næss E, Kristjansson K. Analysis of a vertical flat heat pipe using potassium working fluid and a wick of compressed nickel foam. Energies. 2016; **9**(3):170

[51] Ramanauskas V, Maziukienė M, Miliauskas G. The combined heat and mass transit processes of water droplets in biofuel technologies. Energetika. 2017; **63**(2)

[52] Jonynas R, Puida E, Poškas R, Paukštaitis L, Jouhara H, Gudzinskas J, et al. Renewables for district heating: The case of Lithuania. Energy. 2020;**211**: 119064

[53] Iliopoulos I, Karampekios A, Pandis P, Vourdas N, Jouhara H, Tassou S, et al. Evaluation of organic coatings for corrosion protection of condensing economizers. Procedia Structural Integrity. 2018;**10**:295-302

[54] Pandis PK, Papaioannou S, Siaperas V, Terzopoulos A, Stathopoulos VN. Evaluation of Zn-and Fe-rich organic coatings for corrosion protection and condensation performance on waste heat recovery surfaces. International Journal of Thermofluids. 2020;**3**:100025

[55] Hutny W, Lee G. Improved radiative heat transfer from hydrogen flames. International Journal of Hydrogen Energy. 1991;**16**(1):47-53

[56] Sharma A, Shukla A, Singh R. Low Carbon Energy Supply Technologies and Systems. Taylor & Francis CRC Press; 2020

[57] Matsunaga H, Usuda T, Yanase K, Endo M. Ductility loss in ductile cast iron with internal hydrogen. Metallurgical and Materials Transactions A. 2014; **45**(3):1315-1326

[58] Samanta C. Direct synthesis of hydrogen peroxide from hydrogen and oxygen: An overview of recent developments in the process. Applied Catalysis A: General. 2008;**350**(2): 133-149

Chapter 6

Refractories for Ammonia Production in Fertilizer Unit

Indra Nath Chakraborty

Abstract

Apart from being used as a fuel, natural gas is used extensively for production of ammonia-based fertilizers. During the process of ammonia production natural gas is steam reformed for the production of Hydrogen and the same is converted into Ammonia, by Haber's process, using nitrogen from air. Refractories are required for reformer lining since they are operated at high temperatures as well as in corrosive gas, primarily Carbon Monoxide and Hydrogen, environment. The refractories selected for reformer, thus, should resist the reformer operating temperature as well as the aforementioned gases. Owing to the presence of steam in the working environment magnesia and lime-based basic refractories cannot be used owing to their hydration tendency and thus, aluminosilicate refractories are the only choice. The effect of H_2 and CO on aluminosilicate refractory is the primary focus of this paper. The main concerns are the reduction of siliceous components of the refractory by hydrogen and carbon deposition due to Carbon monoxide decomposition by Boudouard reaction. The effect of these gases on aluminosilicate refractories have been reviewed and based on the outcome suitable refractories have been recommended for ammonia production.

Keywords: ammonia, fertilizer, steam reform, natural gas Haber's process, hydrogen, carbon monoxide, steam, Boudouard reaction, aluminosilicate refractories

1. Introduction

Ammonia is one of the major raw materials for fertilizer industries. It is used for the production of nitro compounds, urea, ammonium sulfate, ammonium phosphate etc. The best and cheapest available source of nitrogen, for ammonia production, is air. Whereas hydrogen, required for ammonia production, is produced from various feedstocks but currently it is derived mostly from fossil fuels. Natural gas, Naphtha, Fuel oil, coal is used for the production of hydrogen. Natural gas is the most preferred option since it is the cheapest in terms of relative investment as well as relative specific energy requirements for ammonia production.

Any industrial process involving high temperature requires refractory. Since the production of Ammonia from natural gas, Ammonia-based fertilizer as well as its derivative involves high temperature, their production, thus, also requires refractory.

The literal meaning of refractory is "Stubborn." In the context of industrial processes, refractories connote the materials which are not markedly affected by their environment. In other words, refractories retain their original features as well as characteristics in aggressive industrial process conditions. During their usage, refractories are exposed to:

- Chemical attack, that is, interaction with the liquid as well as gases
- Abrading action
- Periodical heating and cooling cycle leading to mechanical stress, and/or
- Continuous high temperature, etc.

Against this backdrop the selected refractories, for any industrial process, should ensure that they do not undergo chemical degradation and also can withstand abrasive actions and mechanical as well as thermal stresses of the environment. In this context it also is prudent to mention that no refractory is everlasting. The primary objective is to select a refractory, for any industrial process, such that the impact of the aforementioned industrial process conditions is minimal and the refractory life is maximized. Performance of the refractory is one of the major determinants of the economic efficiency of virtually all industrial processes.

In this context it should also be mentioned that all the aforementioned refractory wearing parameters are not important or significant for all the industrial processes and as a consequence all the refractory properties are not relevant for a given process environment. For the requisite or best performance, the refractory wear contributors, for a given industrial process, need to be identified and refractories should be selected such that it can withstand the identified critical wear parameters. The relevant refractory properties, which would counter the critical wear contributors, need to be optimized. Identification of critical wear parameters, hence, is one of the most important steps for maximization of refractory life at the lowest cost. Apart from the refractory properties, operating parameters of the industrial units also play a key role in determining the refractory performance.

In this paper a brief production process of Ammonia, from natural gas, would be presented (**Figure 1**) and the emphasis would be on the refractory selection process for the ammonia production unit. The primary focus of this discussion would be to understand the operating condition at each step of ammonia production. Once the operating conditions are identified, rationale behind recommending refractories for the various units of the ammonia production process would be discussed. Needless to mention, the primary objective of this paper is to correlate the operating conditions with refractory properties, not the impact of process parameters on efficiency of Hydrogen and thus, ammonia production. For improving the process efficiency catalysts are used at all stages of hydrogen production as depicted in **Figure 1**. Catalysts and their impact on the Hydrogen / Ammonia production process is also not part of this paper. It is evident from the flow diagram (**Figure 1**) that at different stages of the process, part of the hydrogen yield is used as reactant. Prior to venturing into the hydrogen production process, basic information on refractory material would be shared and this would set the backdrop of the rationale behind the refractory recommendations for hydrogen/Ammonia production from natural gas.

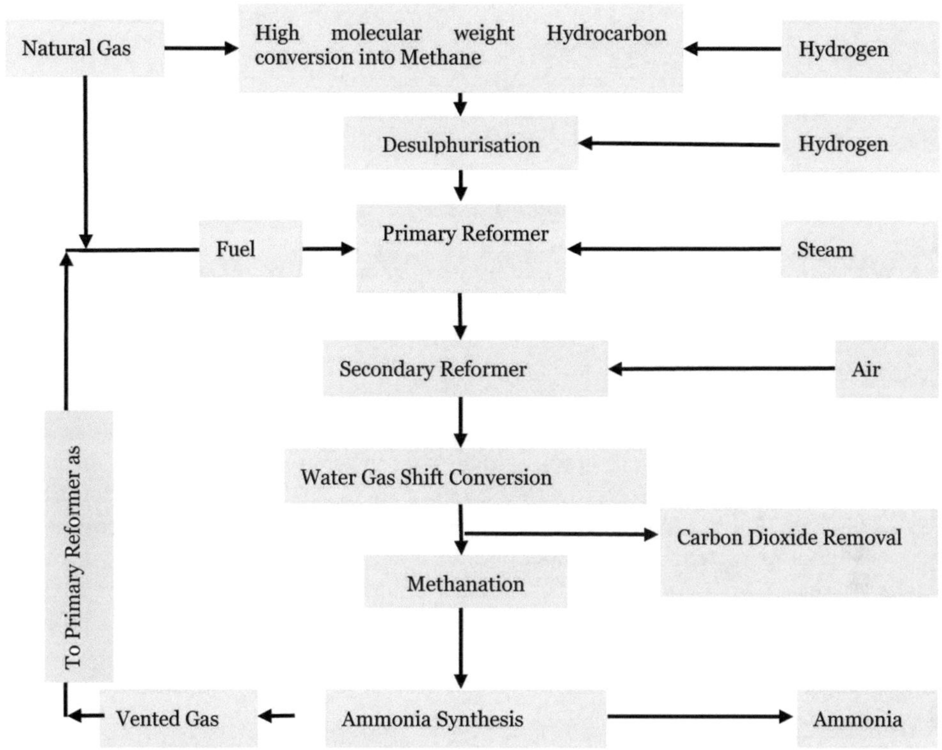

Figure 1.
Process diagram for ammonia production.

1.1 Refractory

Chemically, refractories can be classified as Acidic, Basic and Neutral. The chosen refractory should be compatible with the chemical environment of the process equipment. For example, in case the environment of a process equipment, during operation, is chemically basic, a basic refractory needs to be selected for its lining. Based on the shape, refractory can be classified as shaped, which colloquially is known as bricks, and monolithic refractories. Unlike bricks, monolithics do not come with any specific shape. Shaping of the monolithic materials is done during installation as per the contour of the process equipment. A large number of refractory products fall in the monolithics category. Castable, which is akin to concrete with different material bases and cement, constitutes the largest volume of monolithics. Apart from castables there are numerous monolithic materials, which are installed mechanically, that is, their usage enhances the refractory installation rate, and thus, assists in reduction of process equipment downtime.

Monolithic refractories, in general, have the following advantages over shaped ones.

- Monolithic refractories do not require shaping or firing, and thus, can be produced in a very short notice

- Any shape can be given to monolithics during their installation, i. e. high shaping flexibility

- Since no firing is involved for monolithic production, energy requirement for their production, hence, is lower
- Since the installation of monolithics can be mechanized, their installation rate is significantly faster
- Mechanized installation of monolithics eliminates human error
- With the developments on the material front, monolithic refractories virtually have the same chemistry as bricks. This implies monolithics can be used in all industrial applications with the same operational efficiency as well as performance of bricks

Morphological features of refractories are schematically illustrated in **Figure 2**. All classes of refractories consist of granular material of various sizes, which are termed as aggregates and they are bonded together by very small sized material, which is designated as matrix. Refractories can be conceptualized as the matrix being the continuous phase, where the aggregates are embedded. In majority of the cases, aggregate and matrix chemistry are very different from each other. This makes refractories a heterogeneous material and thus, more complex. Apart from aggregate and matrix, pores are an inherent constituent of the refractories. Refractory porosity and its size distribution can be varied by controlling the proportion of aggregate of different sizes used in the refractory formulation.

Figure 3 illustrates different kinds of pores present in the refractories. Channel pores are the ones which are open from both the ends, though the connecting path may be tortuous. Open pores are the ones which are open from one end but closed from the other. Closed pore, as the name suggests, is closed from all around, i. e. it is not accessible by any gas or liquid in contact with the refractory.

Pore concentration as well as its size distribution in the refractories virtually governs its mechanical as well as thermal properties. For example, channel pores are the only ones which contribute to gas permeation. As the combined concentration of channel and open pores grows higher, the vulnerability of the refractories to chemical attack increases. Total concentration of all 3 types of pores determine the thermal conductivity as well as the strength. Pore concentration of refractories also determines their abrasion resistance, elasticity etc.

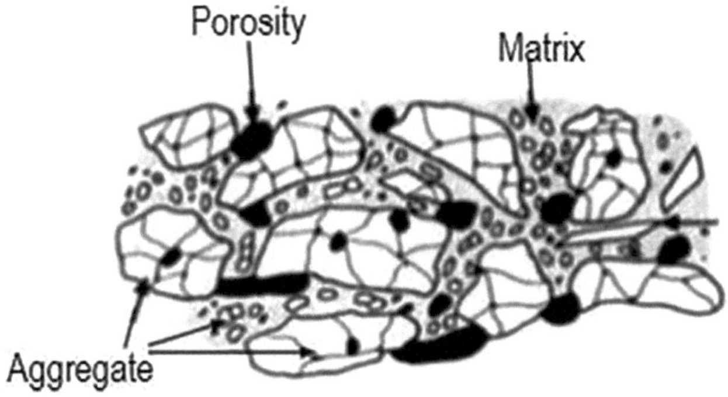

Figure 2.
A schematic representation of refractory showing the heterogeneity of the product.

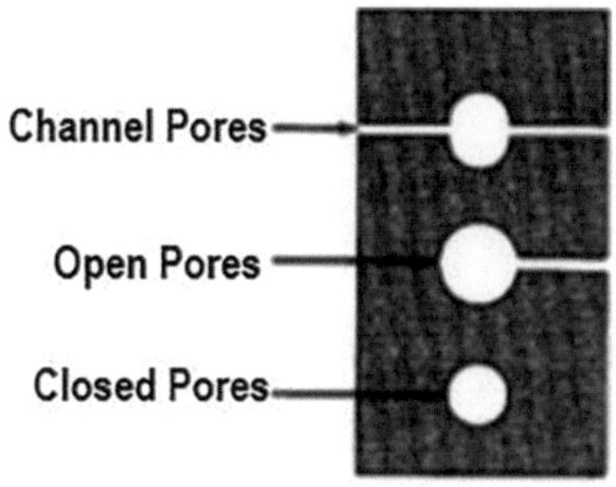

Figure 3.
Schematic representation of types of pores in a refractory.

Selected refractories for any industrial application should conform to the requisite pore structure, its concentration as well as size. For example, for low thermal conductivity, pore concentration of the refractories should be high. On the contrary, for high strength, good resistance to abrasion as well as chemical attack, the refractory porosity should be low. So, not only the chemical compatibility of the refractories with the operating environment ensures the desired performance but its morphological features also play a significant role.

It is, thus, prudent that against the backdrop of the operating conditions of each of the units for Ammonia production, refractory recommendation is made. In the next section, hence, the Ammonia production process as well as the influence of the process parameters on the process efficiency would be discussed in brief. This will set the tone as well as backdrop of refractory selection for the fertilizer industry, particularly for ammonia production units.

2. Natural gas conversion into ammonia

No dearth of information is available in the literature for commercial ammonia production by Haber - Bosch process [1–14]. Among fossil fuels, Natural gas is the most preferred option for Hydrogen generation for Ammonia production. It is well known that the major constituent of Natural Gas is Methane. Apart from Methane, natural gas also contains hydrocarbons of higher molecular weight. Higher hydrocarbons are usually converted into methane by hydrocracking (Eq. 1) prior to further processing.

$$C_nH_m + \{(4n - m)/2\}H_2 = nCH_4 \quad (1)$$

The hydrocracking reactions are endothermic. The reactions are carried out at (65–140 bar) and (400–800°C), in the presence of hydrogen.

The subsequent step for hydrogen generation is Sulfur removal from natural gas. Sulfur bearing compounds in natural gas need to be removed since it deactivates the catalysts used in the Ammonia production process. Sulfur, which is present in the natural gas as Thiol – Sulfur (RSH), is removed by catalytic hydrogenation and in the process hydrogen sulfide is generated (Eq. 2).

$$H_2 + RSH \rightleftarrows RH + H_2S\ (gas) \quad (2)$$

Hydrodesulfurization reaction occurs at 300 to 400°C and 30 to 130 bar absolute pressure. H_2S generated by this process is passed through beds of zinc oxide yielding zinc sulfide (Eq. 3).

$$H_2S + ZnO \rightleftarrows ZnS + H_2O \tag{3}$$

Desulphurized methane is treated with high-temperature steam (700–1000°C) at 3–25 bar pressure, in the presence of a Nickel catalyst in Primary Reformer. The following reactions (Eqs. 4 and 5) occur during interaction of Methane with steam.

$$CH_4 + H_2O \rightleftarrows CO + 3H_2 \ \Delta H = +206.1\ kJ/mol \tag{4}$$

$$CH_4 + 2H_2O \rightleftarrows CO_2 + 4H_2 \ \Delta H = +165.0\ kJ/mol \tag{5}$$

All the reforming reactions are endothermic. The primary objective of the steam reforming process is to maximize Hydrogen yield. CO and CO_2 are the byproducts of the steam reforming process. Apart from the steam reforming reactions, Dry Reforming and Water Gas Shift (WGS) reactions (Eqs. 6 and 7) also occur by virtue of interactions between the steam reforming reaction products, viz. CO and CO_2, steam and Methane.

$$CH_4 + CO_2 \rightleftarrows 2CO + 2H_2 \qquad \Delta H = 247.3\ kJ/mol \tag{6}$$

$$CO + H_2O \rightleftarrows CO_2 + H_2 \qquad \Delta H = -41.15\ kJ/mol \tag{7}$$

WGS is called so since by virtue of this reaction the ratio of Water Gas constituents, viz. CO, and H_2, are altered. In this specific case, the reaction is carried out in such a way that the reaction proceeds in favor of Hydrogen generation. The yield of Primary reformers typically contains 60% Hydrogen. All the reactions in Primary Reformer are reversible in nature. Owing to this reason pressure, temperature, and ratio of the reactants determine the extent of Hydrogen generation. **Figures 4** and **5** illustrate the impact of Steam - Methane ratio, temperature, and pressure on the reforming process. The observations are in the expected line of the reversible reactions and thus, the conversion of natural gas into CO and H_2 increases with:

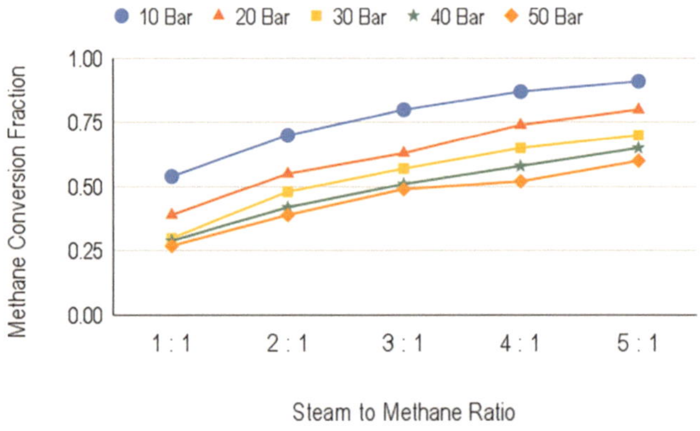

Figure 4.
Effect of pressure on the steam reforming process in primary reformer for the operating temperature of 800°C [6].

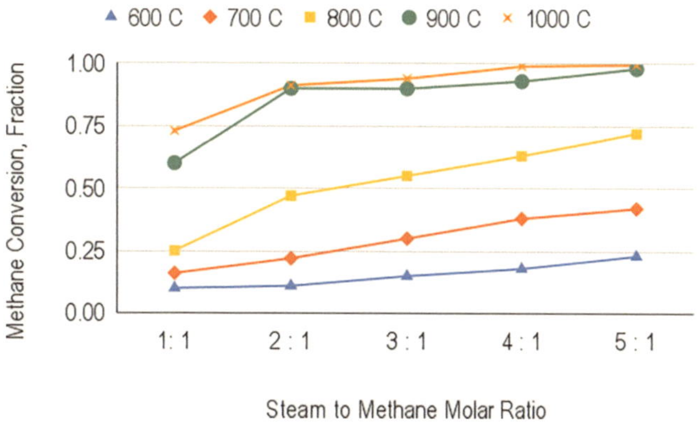

Figure 5.
Effect of temperature on the steam reforming process in primary reformer for operating pressure of 30 Bar [6].

- Increase in temperature
- Increase in steam to methane ratio, and
- Decrease in pressure

In this context it is prudent to mention that Nickel catalyst, which is used for steam reforming reaction, also activates Boudouard reaction (Eq. 8) [6]. Since this reaction is reversible in nature, higher pressure favors the reaction to proceed to the right, that is, higher pressure favors CO decomposition.

$$2CO \rightleftarrows C + CO_2 \tag{8}$$

By virtue of this reaction Carbon deposition tendency is fairly severe. This reaction has a significant impact on the life of refractories as well as its performance, which would be discussed in the refractory selection section.

The yield of theprimary reformer, that is, Hydrogen (H_2), unreacted CH_4, unreacted water (steam), CO, CO_2, Nitrogen, Argon, etc. are fed in the secondary reformer, via Transfer Line, for further processing to increase the Hydrogen yield. Air, apart from the yield of the Primary reformer, is a feed for the secondary reformer. The primary objective of the Secondary reformer is to complete Steam Reforming as well as WGS reactions to produce further quantities of Hydrogen and adjust the hydrogen and nitrogen ratio for Ammonia production.

Apart from the continuation of steam reforming, dry reforming, and WGS reactions (Reactions 4–7), the following reactions also take place in Secondary Reformers.

$$2CH_4 + O_2 = 2CO + 4H_2 \tag{9}$$

$$2CH_4 + 3O_2 = 2CO + 4H_2O \tag{10}$$

$$CH_4 + 2O_2 = CO_2 + 2H_2O \tag{11}$$

In addition to these reactions combustion of Hydrogen as well as Carbon Monoxide also occurs. The exothermic reactions raise the temperature of gases in the secondary reformer to ~1000°C and the conversion to hydrogen achieved is of the order of 99% and the rest is a mix of CO, CO_2, CH_4, and water in chemical equilibrium.

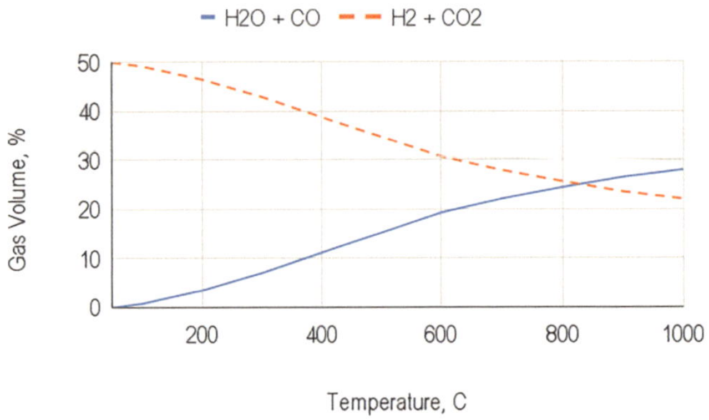

Figure 6.
Effect of temperature on water gas shift (WGS) reaction equilibrium [9].

The yield of the secondary reformer is further treated in the WGS reactor to increase the Hydrogen concentration (Eq. 7). Owing to the exothermic nature of the WGS reaction, lower temperature favors the formation of Hydrogen (**Figure 6**). The yield of secondary reformers is processed in 2 stages, viz. High-temperature WGS reactor at 300–450°C and Low - temperature WGS reactor at 200–250°C.

All oxygen-containing substances, including water, are poisons for the ammonia synthesis catalyst and thus, need to be removed. The gas mix from the WGS reactor, hence, is further processed, for removal of CO and CO_2 by Pressure Swing Adsorption (PSA) or Cryogenic Distillation (CD). After the bulk removal of CO by WGS reaction and CO_2 removal by PSA or CD, the typical synthesis gas still contains 0.2–0.5 vol % CO and 0.005–0.2 vol % CO_2.

Methanation (Eqs. 12 and 13) is the simplest method to reduce the concentrations of the carbon oxides well below 10 ppm and is widely used in ammonia production units. There are two main purposes for methanation, viz. to purify synthesis gas, i. e. remove traces of carbon oxides, and to manufacture methane. CO and CO_2 methanation is carried out in the temperature range of 200–600°C and 350–450°C, respectively.

$$CO + 3H_2 \rightleftarrows CH_4 + H_2O \quad (12)$$

$$CO_2 + 4H_2 \rightleftarrows CH_4 + 2H_2O \quad (13)$$

The hydrogen - nitrogen mix, obtained after methanation, is used for Ammonia production by the Haber - Bosch process.

3. Refractories of ammonia production

It is apparent from the previous section that the environment of all the units for ammonia production contains steam, since it is one of the reactants used in ammonia production. MgO or CaO based, that is, basic, refractories, thus, would be unsuitable since they are prone to hydration. The hydration of both CaO and MgO yields hydroxides. The formation of these oxides is accompanied by large volume expansion. As a result, the soundness of the basic refractories in the ammonia production unit would be destroyed and thus, are unsuitable.

Against this backdrop, it can be concluded that chemically only aluminosilicate refractories would be suitable. Aluminosilicate refractories, as the name suggests, are based on Alumina (Al_2O_3) and Silica (SiO_2). One compound, viz. Mullite ($3Al_2O_3.2SiO_2$), which contains 72% Al_2O_3, is the only thermally stable binary phase in the aluminosilicate system. With the increase of alumina content of aluminosilicate refractories, their refractoriness, i. e. temperature withstanding capability, increases. Apart from alumina and silica, aluminosilicate refractories also contain certain impurities, which are inherently present in the natural aluminosilicate raw materials used for their production. Primary impurities in aluminosilicate refractories are iron oxide, titanium dioxide, alkali, and alkaline earth oxides and these minor constituents play a significant role in deciding the ultimate refractory performance in any given environment and hence, the impact of these impurities also should be evaluated.

3.1 Impact of hydrogen on aluminosilicate refractories

The effect of Hydrogen gas on fused silica and a wide range of aluminosilicate refractories has been studied in depth [15–20]. The primary effect is reduction of silica by hydrogen gas as per Eq. 14 [16].

$$SiO_2\ (s) + H_2\ (g) \underset{\text{Lower Temperature}}{\overset{\text{Higher Temperature}}{\rightleftarrows}} SiO\ (g) + H_2O\ (g) \quad (14)$$

The gaseous products generated by virtue of reaction 14 are carried off by the process stream. Downstream, when the temperature is conducive for solidification of SiO, it condenses and gets deposited as SiO_2 and Si mix causing heat-exchanger fouling and product contamination. **Figure 7** illustrates the effect of time as well as temperature on the reduction of fused silica by Hydrogen gas. As expected, with the increase in duration of fused silica and hydrogen interaction the loss of silica increases. The effect is similar when the interaction temperature increases.

The effect of time and temperature in the hydrogen environment on aluminosilicate refractories of different silica concentration follow the similar trend. **Figure 8**

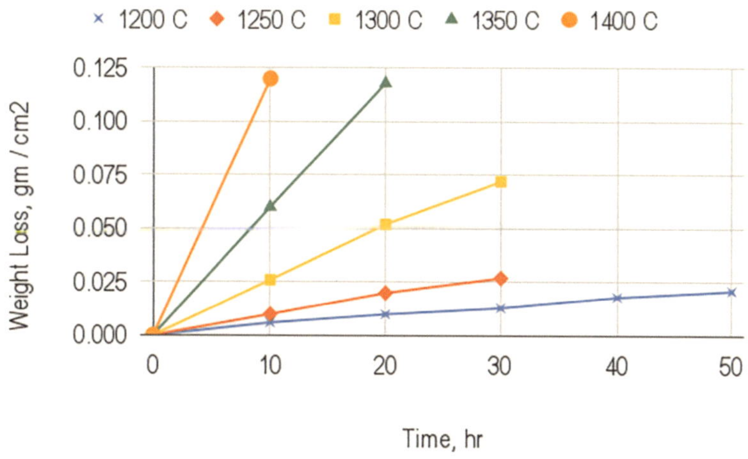

Figure 7.
Effect of time and temperature on fused silica reduction [15].

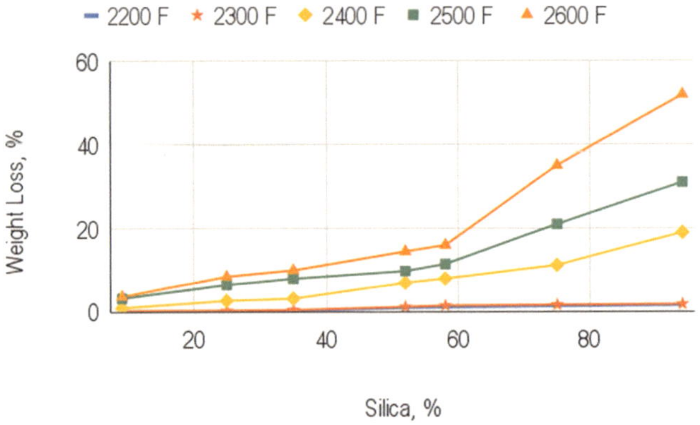

Figure 8.
Effect of temperature on reduction of refractories with different silica content after reduction for 33 hrs in 100% H_2 atmosphere [16].

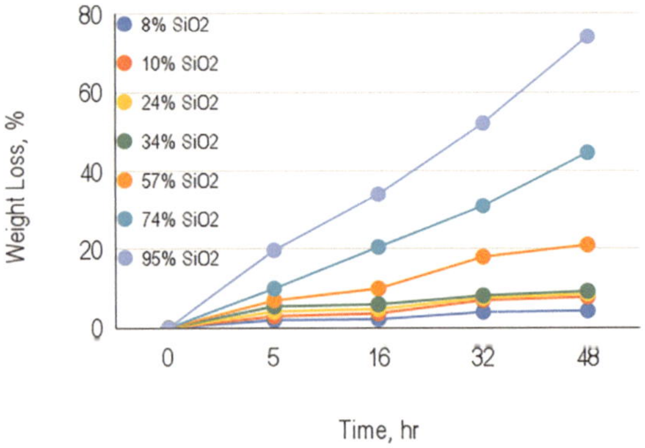

Figure 9.
Effect of time on reduction of refractories with different silica content for reduction by hydrogen at 2600°F in 100% H_2 atmosphere [17].

illustrates that for a given level of silica in aluminosilicate refractories, increase of temperature causes higher loss of silica. It is also seen that for a given temperature, silica loss increases with the increase of silica concentration in aluminosilicate refractories.

Figure 9 illustrates the effect of time on the silica loss of aluminosilicate refractories with different silica content. As contemplated, the increase of refractory - hydrogen interaction duration causes greater silica loss. It also is in the expected line that for a given refractory - hydrogen interaction duration, silica loss increases with its silica concentration. But it is observed that upto 10% silica concentration in the refractory, the silica loss by SiO_2 reduction is marginal.

Figure 10 reports the impact of pressure as well as silica concentration of refractories in the hydrogen environment at 2400°F. Loss of silica, for all levels of pressure, increases with increase of silica concentration. It is fairly evident from the results that for a given silica concentration, the silica loss decreases with the increase of hydrogen gas pressure. In other words, higher operating pressure would protect the siliceous part of aluminosilicate refractories from the Hydrogen present in reformers. The operating pressure of primary as well as secondary reformers is >30 bar and thus, the silica loss from the aluminosilicate refractories is expected to be low.

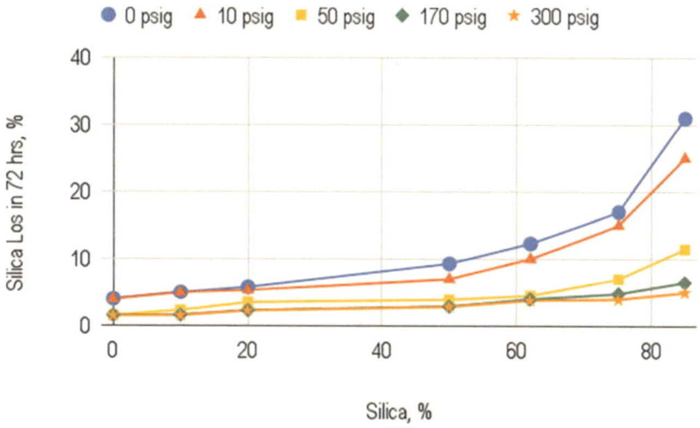

Figure 10.
Effect of hydrogen pressure on reduction of refractories with different silica content. These data are for 2400°F and hydrogen flow of 4.6 liters/minute [17].

3.2 Impact of carbon monoxide on aluminosilicate refractories

The effect of Carbon Monoxide (CO) on the refractories is attributed to its decomposition as per the reverse Boudouard/Bell reaction (Eq. 8) [21–28]. In the context of refractories, it can be stated that CO gas diffuses into open as well as channel pores. By virtue of the reverse Boudouard reaction, carbon is deposited in the pores. Deposited carbon grows and generates stress within the refractories. When the stress exceeds the strength of the refractories, cracks form and the refractories get damaged. In the worst possible scenario, the refractory is destroyed. **Figure 11** illustrates the stability of CO as a function of temperature. With the increase in temperature, the stability of CO increases. It is apparent from the figure that destruction of refractories, by deposition of carbon, will not occur when the temperature exceeds 1100°C and at <500°C, though CO decomposition is thermodynamically possible, the reaction kinetics is slow and hence, CO decomposition rate is low. Refractories, thus, are vulnerable to destruction by CO in the 500–700°C range.

Metallic iron is a known catalyst for CO decomposition [27, 28]. The CO decomposition process proceeds by reduction of iron oxides by CO, to metallic iron. The iron

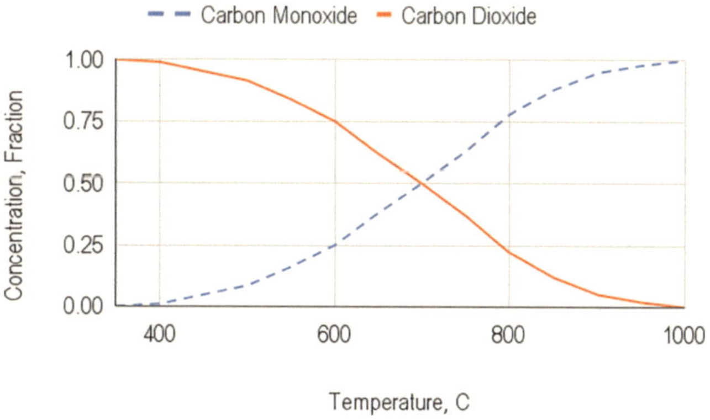

Figure 11.
CO - CO_2 equilibrium as per FACTSage software [21].

subsequently is carburized into Fe - carbide, which decomposes into Fe according to Eq. 15.

$$Fe_3C = 3Fe + C \tag{15}$$

The Fe, formed via Iron Carbide decomposition, has high surface activity and hence, enhances the CO decomposition rate. The catalytic activity is believed to proceed via the formation of Iron Carbide (Fe_3C).

3.3 Impact of simultaneous presence of hydrogen and CO on aluminosilicate refractories

Apart from reverse Boudouard reaction, reverse water gas reaction also contributes to carbon deposition in refractories as per Eq. 16. This reaction, however, can proceed only below 680°C [29]. Beyond this temperature Carbon monoxide and Hydrogen are stable phases and hence, carbon deposition within the refractory is not expected. This prediction, however, is based on the assumption of standard states, i. e. the activity of the reactants and the products are 1. In real situations the reactant as well as product activities would be less than 1, which means there would be certain deviations in the temperature predicted above.

$$CO + H_2 \rightleftarrows C + H_2O \tag{16}$$

Simultaneous presence of H_2 and CO, which is the case for the reformers, enhances the CO decomposition rate when the temperature exceeds 577°C [23]. The presence of H_2, in the CO environment, not only alters the carbon deposition rate but also determines the temperature at which the maximum carbon deposition occurs. For example, for 0.8% and 19.9% H_2, maximum Carbon deposition occurs at 530 and 630°C, respectively. At lower temperature, however, the effect of hydrogen is marginal [28]. It, thus, is a precondition that refractories for the CO environment should be low in their iron oxide content.

3.4 Impact of steam on aluminosilicate refractories

Steam, in general, appears to be inter towards the oxide refractory materials. Steam, however, interacts with silica as per the Eq. 17, which is reversible in nature. It is evident from this equation that SiO_2 in the presence of steam is converted into $Si(OH)_4$.

$$SiO_{2(s)} + 2H_2O_{(g)} \rightleftarrows Si(OH)_{4(g)} \tag{17}$$

Figure 12 illustrates the effect of temperature as well as time, on the weight loss of cristobalite, at 0.84-atmosphere steam pressure. The magnitude of silica loss, via Si $(OH)_4$ formation, increases with the increase of temperature up to 1350°C. Beyond 1350°C, however, the rate, as well as the magnitude of silica loss, reduces. This observation has been attributed to the reversal of Eq. 17. The opinion related to the temperature impact on $Si(OH)_4$ formation in aluminosilicate refractories, however, appears to be different from that for pure silica. It is also believed that at <980°C steam reacts with siliceous components of aluminosilicate refractories and yields Si $(OH)_4$. At >1000°C, however, the same reactants primarily yield SiO[19].

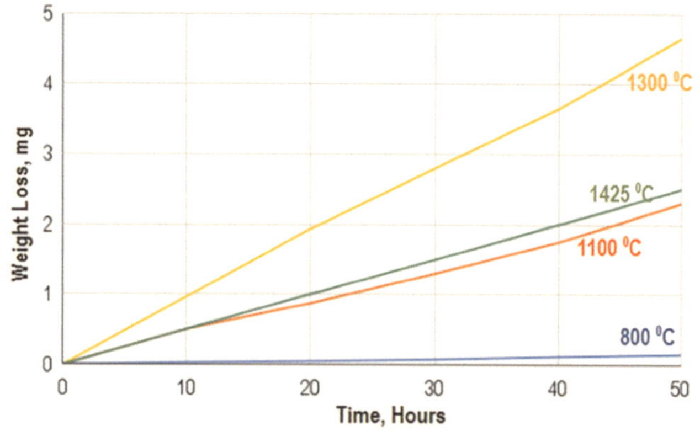

Figure 12.
Effect of steam temperature on the loss of silica by 0.84-atmosphere pressure [17].

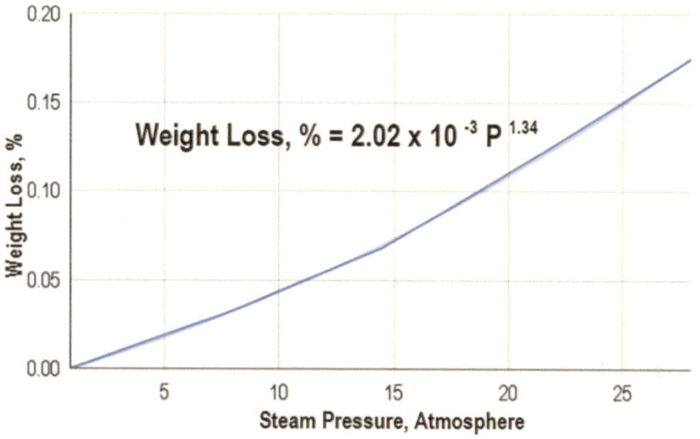

Figure 13.
Effect of steam pressure on the volatilization of - 325 mesh 24 gm Cristabolite in 2 hours at 900°C. the steam flow rate was maintained at 2.73 cm/second [17].

Figure 13 illustrates the impact of steam pressure on cristobalite weight loss by virtue of Eq. 17. It is obvious from the figure that with the increase of pressure, the silica loss increases. The silica loss is related to pressure by Eq. 18 [17].

$$\text{Weight Loss, \%} = 2.02 \times 10^{-3}\ P^{1.34} \tag{18}$$

The index of pressure in Eq. 18 is >1 but <2. Indices of 1 and 2 indicate the reaction is controlled by transportation through the boundary layer and reaction at the interface, respectively. The mechanism of $Si(OH)_4$ formation, by virtue of Silica - steam interaction (Eq. 17), thus, is not very clearly defined.

Figure 14 illustrates the impact of steam, on the weight loss of silica in alumino-silicate refractories, in the hydrogen environment. As has been observed earlier, the weight loss increases with the increase in silica content of the refractories. When steam is present in the environment, along with hydrogen, the rate of silica reduction is reduced significantly. Additionally, silica loss shows only marginal dependence on the silica content of the refractory, when steam is present in the hydrogen

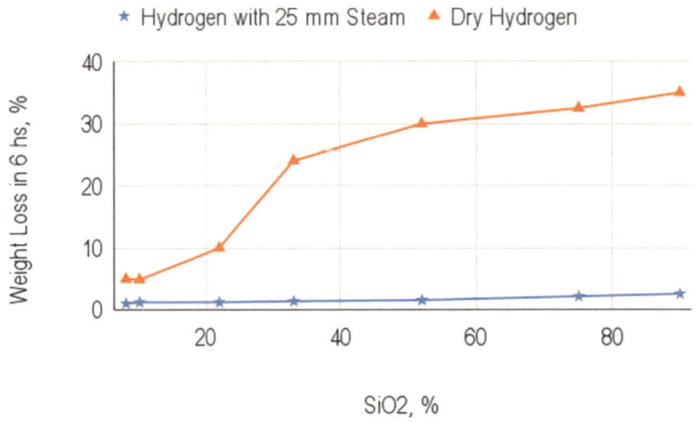

Figure 14.
Effect of steam on reduction of refractories with different silica content in a hydrogen atmosphere at 2400°F [18].

environment. In short, the impact of hydrogen on silica reduction is reduced significantly when steam and hydrogen are present simultaneously. This implies that the presence of steam in the working environment, which will be the case in both the reformers, would protect the aluminosilicate refractories better by reducing the reductant effect of hydrogen.

3.5 Impact of simultaneous presence of hydrogen and steam on aluminosilicate refractories

Figure 15 illustrates the impact of the simultaneous presence of Hydrogen and steam on silica loss of various aluminosilicate refractories. It is evident from the figure that the silica loss in the hydrogen atmosphere comes to a rest, for the aluminosilicate refractories, in the presence of steam. In other words, the reduction of silica by hydrogen ceases when Hydrogen and steam are present simultaneously. The corollary of the same is the adverse impact of steam, on aluminosilicate refractories, is annulled by the presence of hydrogen.

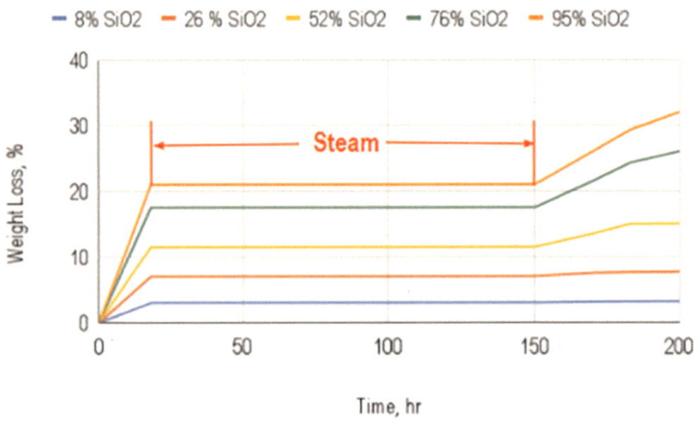

Figure 15.
Impact of steam on the reduction of aluminosilicate refractories at 2500°F by hydrogen [18].

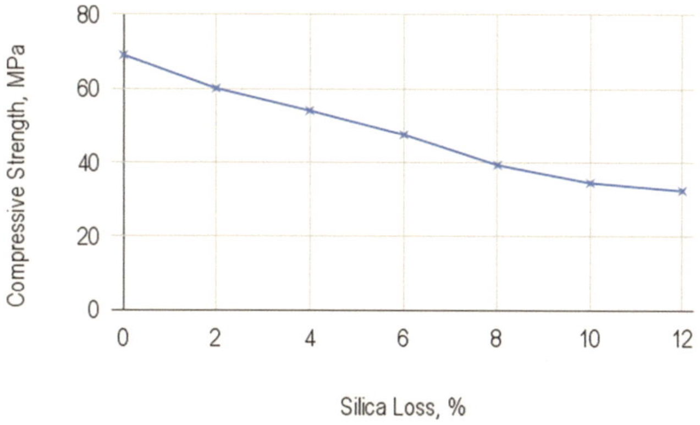

Figure 16.
Effect of silica loss on strength of refractories. The reduction was carried out in the hydrogen atmosphere [16].

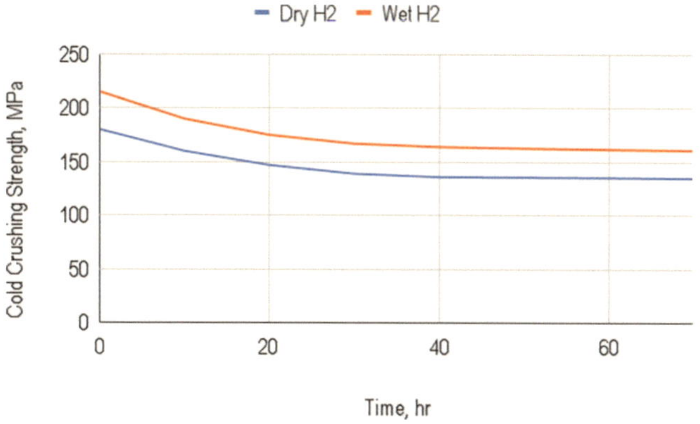

Figure 17.
Impact of 25 mm steam on strength reduction of 85% alumina brick in a hydrogen atmosphere at 2500°F [16].

3.6 Impact of gaseous environment on the refractory characteristics

Reduction of the siliceous component, by hydrogen, increases the refractory porosity and increased porosity adversely affects the refractory strength [15] **Figure 16** illustrates the impact of silica loss on the strength of 52% alumina bricks. As expected, the increase in silica loss leads to a greater loss of strength. A loss of 10% silica causes approximately 50% strength reduction.

As has been seen in the earlier section, the reduction of silica by hydrogen is inhibited by the presence of steam. The reduction of refractory strength, hence, is lesser in the presence of steam, compared to when Hydrogen is present by itself (**Figure 17**). It is evident from the illustration that the reduction of strength for an aluminosilicate brick is lower by approximately 20% when the hydrogen environment contains steam. In fact, during ammonia production steam is always present in the operating environment of all the units. The impact of hydrogen, as a reductant thus, is expected to be lower in all the units of the ammonia production process.

Figure 18 illustrates the impact of alumina as well as iron content of the refractories in the CO environment. It is evident that CO by itself gets decomposed into C and CO_2 as per the reverse Boudouard (Bell) reaction (Eq. 8). This is reflected by

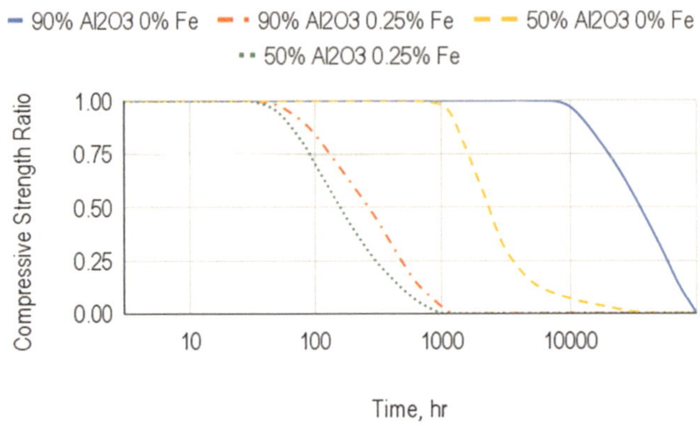

Figure 18.
Impact of 1 Bar carbon monoxide atmosphere on strength for refractories with different alumina and iron content [21].

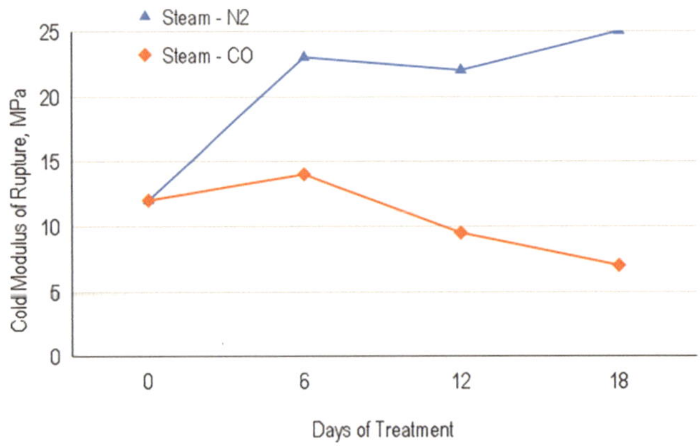

Figure 19.
Effect of CO - steam on the strength of WTA based Castable at 32 bar pressure [26].

the reduction of the strength of Fe - free 45 as well as 90% alumina containing refractories with an increase in interaction time with CO. It also is obvious from **Figure 15** that Fe acts as a catalyst for the decomposition of CO. In the presence as well as the absence of iron in the refractories, the strength reduction rate of 50% alumina refractory is faster, compared to the one containing 90% alumina [22]. 50% Alumina refractory is destroyed approximately 5 times faster than 90% alumina products. Lower alumina refractories, thus, are more vulnerable to disintegration due to CO decomposition.

Literature reports enhancement of CO disintegration in the presence of steam [26]. **Figure 19** illustrates that the strength of White Tabular Alumina (WTA) based castable reduces in CO - Steam atmosphere. On the other hand, it is observed that the presence of Nitrogen, in the CO environment, enhances the strength of WTA-based castable, that is, the trend is reversed when Nitrogen is replaced by steam. Nitrogen is an inert gas and by itself, it does not react with aluminosilicate refractories. But when it is present together with CO gas, it alters the interaction process with the refractories and enhances its strength. In fact, Nitrogen is present in the operating environment in secondary reformers and downstream. The adverse effect of steam on castable strength, in the CO environment of the ammonia production unit, is nullified to some extent owing to the simultaneous presence of nitrogen.

3.7 Ammonia plant operating condition

The operating conditions of ammonia production units can be summed up as follows:

- The gases present in the working environment of all the units are Hydrogen, CO, CO_2, Steam, Nitrogen, and Methane.
- Hydrogen concentration increases downstream, whereas the same for CO, CO_2, steam, and CH_4 decreases.
- Typical operating temperature and pressure of primary reformer is 980°C and 35 bar, respectively.
- Whereas the same for secondary reformers is 1350°C and 35 bar, respectively. Such high temperature in secondary reformers is observed only around the combustion region. The temperature of the other sections of secondary reformer typically is 1100°C
- Operating temperature and pressure of Transfer Line, which allows the passage of primary reformer yield to secondary reformer, are in the vicinity of those in the reformers
- The high temperature shift converter operates at 480°C, whereas the operating temperature of low temperature shift converter is 230°C
- Methanator usually operates at $<$350°C
- The typical operating temperature of ammonia production unit, thus, does not exceed 1100°C, except around the burner in secondary reformer, where the temperature is $\sim$1350°C
- Operating temperature of each unit in the Ammonia production plant is different. The operating temperature of a given unit, however, does not fluctuate. Additionally, the operation is continuous. These 2 aspects imply that no temperature fluctuation occurs in a given unit of ammonia production plant, that is, there is no thermal shock on the refractories.
- Processing units contain only gases, that is, they contain no solid or liquid. The abrasion experienced by the refractories, thus, is minimal. Abrasion, hence, is not a major issue for the refractories in ammonia production units.
- Except for the dome section of the secondary reformer, which supports the catalyst carriers, there is no direct load on the refractories in any of the units

3.8 Refractories for ammonia production unit

As per conventional wisdom, the ammonia production unit operating temperature is not a matter of concern since it is moderate and does not exceed 1350°C. Considering only the operating temperature, hence, it can be stated that even the lowest grade of aluminosilicate refractories, except for the secondary reformer dome, would suffice. Only the strength of the refractories should be high enough to withstand the operating pressure. But is this the case when we consider the gaseous environment in ammonia production units?

The discussion in the previous sections revealed that even the "inert" gases, like nitrogen and steam, influence the refractory properties and hence, performance. It also is evident that it is not only the high temperature that damages the refractories. Rather for certain working environments, for example, destruction of refractories by CO, low, rather than high, temperature causes greater damage to refractories. It is not only operating temperature, but the impact of the gaseous environment also needs to be considered for refractory selection.

The gases from the working environment can permeate through channels as well as open pores (**Figure 3**) to the interior section of the refractories. In the specific case of reformers, where the operating pressure is of the order of 35 bar, gas permeation is a very likely possibility. Additionally, refractories have finite thermal conductivities, i. e. the temperature along with the thickness of refractory lining decreases.

The silica reduction rate, by hydrogen, increases with the increase of temperature and becomes significant when the temperature exceeds 1100°C (**Figure 7**). This implies that silica reduction by hydrogen is expected to occur on the refractory hot face, not in the interior part. Refractory destruction by CO, on the other hand, will not occur on the refractory hot face since the reformer operating temperature is >1000°C. But at a certain distance from the hot face, along with the thickness of the refractory, where the temperature is in the 500 to 700°C range, the extent of CO decomposition and as a consequence refractory destruction would be high.

In short, the refractories remain vulnerable to destruction by CO even if the operating temperature of the unit is >1000°C. In the context of ammonia production units, thus, refractories are at risk of destruction by CO and the destruction process may commence at the interior part of the refractory, where the temperature is conducive for CO decomposition.

During the interaction of aluminosilicate refractories with hydrogen, silica is preferentially attacked and this leaves a porous alumina network [15]. Both Hydrogen, as well as CO, permeate through the residual porous alumina structure and thus, silica reduction by hydrogen as well as CO attack continue in the interior part of the refractory. The SEM analysis of the used refractories from an ammonia plant transfer line shows that the loss of silica occurs from the aggregate and not the matrix [19]. This observation contradicts the conventional belief that the matrix, being a finer part of the refractory formulation, has higher reactivity and thus, is more vulnerable to chemical attack.

Since primary as well as secondary reformers and transfer lines operate at high pressure, at ~35 bar, it is prudent to see the impact of pressure on Hydrogen and CO attack of refractories. An increase in pressure reduces the rate of silica reduction by hydrogen (**Figure 10**). Higher operating pressure, thus, is favorable for the prevention of refractory degradation by silica reduction. Le chatelier principle, on the contrary, predicts that an increase of pressure would enhance the decomposition rate of

CO into carbon (Eq. 8). Such contradicting situations make the refractory selection for ammonia plants more convoluted.

Based on the discussion of the previous section it is obvious that the combined effect of Steam - Nitrogen - CO and H_2 on the aluminosilicate refractory is fairly contradicting as well as complex since:

- Refractory degradation, in the CO environment, is enhanced in the presence of steam or hydrogen and also when they are present simultaneously. Steam on the other hand inhibits the reductant effect of hydrogen.

- Simultaneous presence of CO and Nitrogen, however, reverses the refractory degradation process enhanced by steam. Both steam, as well as nitrogen, are present in the secondary reformer environment

- Silica reduction, by hydrogen, increases with the increase of temperature. On the contrary, CO gas does not decompose into CO_2 and C beyond ~1000°C since it is thermodynamically stable at this temperature.

- Increase in operating pressure reduces the reductant effect of Hydrogen. Whereas an increase in pressure increases the decomposition rate of CO

- Steam, if present by itself, causes silica loss by the way of $Si(OH)_4$ formation up to ~1350°C. Beyond this temperature, however, $Si(OH)_4$ formation is suppressed and loss of silica is prevented. Up to 1350°C silica loss increases with the increase of steam pressure.

- In the presence of Hydrogen, the adverse effect of steam, on aluminosilicate refractories, is suppressed.

Apart from the opposing conditions of refractory-gaseous environment interactions, marginal volatilization of silica from the refractory may poison the catalysts downstream. Refractories, hence, are selected so that even the slightest silica reduction in the process is mitigated. This is the reason why the lowest possible iron oxide as well as silica-containing refractories, are selected for primary as well as secondary reformers. This is because, in the absence of iron, the CO decomposition rate is reduced, whereas the reductant effect of hydrogen is prevented by the absence of silica in the refractories. Needless to mention that such high purity raw materials are not available in nature and thus, the refractories for primary and secondary reformers should be based on synthetic raw materials like WTA or White Fused Alumina (WFA), which are low in both iron oxide as well as silica.

Refractories, based on synthetic raw materials like WTA or WFA, are typically recommended only for locations where the operating temperature exceeds 1650°C. Primary and secondary reformers are examples, where although the operating temperature is low, still synthetic raw material-based refractories are recommended due to the gas composition in their operating environment as well as the possibility of catalyst poisoning. In other units of ammonia making plant, that is, other than primary and secondary reformers, the conditions are not as severe due to the operating temperature being lower than 500°C, which is the temperature of CO decomposition initiation, and hence, lower alumina aluminosilicate refractories with lesser purity would suffice.

4. Brick or monolithic for ammonia plant

Owing to the progress made on the material development front, there is no significant difference between brick and monolithic chemistry. The gap between brick and monolithic has been bridged primarily by reducing the flux concentration in monolithic formulations. For example, the fluxing component in castable and majority of aluminosilicate gunning formulations is CaO originating from the cement. With the progress in materials technology, aluminosilicate castables and gunning materials can be produced without any cement, that is, without CaO. And thus, in the majority of industrial applications, refractory bricks can be replaced by monolithic.

The aforementioned progress in materials technology can be exploited and monolithic, instead of bricks, based on WTA, can be used in ammonia plants. The primary advantage of monolithic, that is, mechanized installation, thus, can be capitalized for the refractory lining of ammonia plants. The fallout of mechanized installation of monolithic is the requirement of skilled bricklayers is eliminated and simultaneously, the refractory installation rate also is enhanced. Monolithics, depending on its type, the installation rate is 2.5 to 10 times higher than those for the bricks. The substitution of bricks by monolithics, with similar chemistry, would reduce the inventory cost, delivery time, dependence on human skill, and above all the installation time. All these factors put together would reduce the plant downtime when a brick is replaced by monolithic.

Figure 20 illustrates the installation rates of various classes of monolithics and their characteristics. Castables and gunning materials typically have 1.5 and 8% CaO, respectively. Apart from the high concentration of CaO, the strength characteristic of the gunning formulations is not very favorable. But the time required for gunning material installation is significantly lower than that for the castable. New generation gunning formulations, which also are known as wet gunning or shotcreting, have strength characteristics as well as CaO concentration similar to that of castable. On the contrary, the time required for shotcrete installation is comparable to that for gunning material. An additional advantage of shotcrete material is extremely low rebound loss, that is, lower dust generation during installation compared to that of gunning material. This makes shotcrete significantly more user as well as environment-friendly

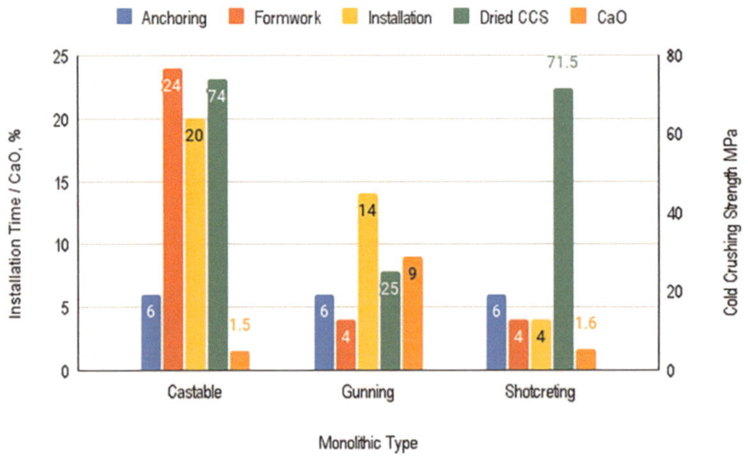

Figure 20.
Features and installation aspects of castable, gunning, and Shotcreting materials.

compared to gunning material. In short, shotcrete materials have the advantages of both castable as well as gunning materials, that is, faster installation like gunning material and installed material property as well as chemistry similar to that of castable.

In short, low iron - low silica-alumina based bricks and monolithics are recommended for primary as well as secondary reformers. Lower alumina products with lesser purity would suffice for the rest of the ammonia plant.

5. Conclusion

Going by conventional perception 45% alumina refractories would suffice for the fertilizer industry since the operating temperature does not exceed 1350°C. This, however, is not the case owing to the simultaneous presence of CO, H_2, CO_2, Steam, and N_2 in the environment of ammonia production units. Generally, steam and N_2 are treated as inert gases in regard to their interaction with refractory materials. But steam reduces the reduction rate of SiO_2 of Aluminosilicate refractories by Hydrogen gas. Steam, on the other hand, enhances the decomposition of CO as per the Boudouard reaction. The favorable effect of steam for silica loss by its reduction, hence, can not be exploited owing to the simultaneous presence of CO and Hydrogen in the ammonia production unit environment. The aluminosilicate refractories recommended for ammonia production units, mainly for primary and secondary reformers, should, thus, be low in iron oxide as well as silica content so that the adverse impact of both Hydrogen and CO is minimized. Such formulation can be either in the form of bricks or monolithic. Refractories with such stringent chemistry requirements can only be met by those based on synthetic raw materials like WTA or WFA.

The fertilizer industry is one good example that epitomizes that the alumina content of aluminosilicate refractories is not decided only by the operating temperature. Owing to the simultaneous presence of Hydrogen, CO, N_2, and steam in the operating environment, there is no option but to use aluminosilicate refractories with a low concentration of iron oxide as well as Silica through the operating temperature of the reformers barely exceeds 1100°C. Based on this analysis it is evident that operating temperature is not the only determinant of refractory quality for an industrial process but the gaseous environment of the unit also plays a significant role in the refractory selection process.

Currently, the reduction of greenhouse gas emissions is a major focus of all industrial processes. To achieve this goal, replacement of fossil fuels by hydrogen for iron production as well as other industrial processes are being targeted. The analyses presented in this paper also will provide direction for refractory selection for the industrial processes where hydrogen is being used or hydrogen is a yield, e.g. Gasification.

Author details

Indra Nath Chakraborty
Calderys India Refractories Limited, Nagpur, India

*Address all correspondence to: indra.chakraborty@imerys.com

References

[1] Anderson JR, Baker BG. The hydrocracking of saturated hydrocarbons over evaporated metal films. Proceedings of the Royal Society of London. Series A: Mathematical and Physical Sciences. 1963;**271**:402-423

[2] Ashour A-DA, Ala'a Abdulrazaq J. Study of chemical reactions effect on the design and the performance of an industrial secondary reformer reactor in the fertilizers plants. International Journal of Applied Sciences and Engineering Research. 2013;**2**:60-69

[3] Ksiazek M, Grådahl S, Eirik RA, Wittgens B. Capturing and condensation of SiO gas from industrial Si furnace. In: Reddy RG, Chaubal P, Pistorius PC, Pal U, editors. Proceedings of the 10th International Conference on Molten Slags, Fluxes and Salts, 22–25 May 2016, Seattle Washington USA. Switzerland: Springer International Publishers; 2016. pp. 1153-1160

[4] Tamião de Campos Roseno Karina, M. B. de Alves Rita, Giudici Reinaldo, Schmal Martin. Syngas production using natural gas from the environmental point of view. In: Biernat K, editor. Biofuels. IntechOpen; 2018. DOI: 10.5772/intechopen.74605

[5] Byron SRJ, Muruganandam L, Shekhar SM. A review of the water gas shift reaction kinetics. International Journal of Chemical Reactor Engineering. 2010;**8**:1-32

[6] Ali MS, Zahangir SM, Badruddoza AZM, Haque MR. A study of effect of pressure, temperature and steam/natural gas ratio on reforming process for ammonia production. Journal of Chemical Engineering. 2010;**23**: 1995-2005

[7] Jörissen L. Residential energy supply: Fuel cells. In: Garche J, Dyer C, Moseley P, Ogumi Z, Rand D, Scrosati B, editors. Encyclopedia of Electrochemical Power Sources. 1st ed. Netherlands: Elsevier; 2009. pp. 108-124. ISBN: 9780444520937

[8] Rice Steven F, Mann DP. Autothermal reforming of natural gas to synthesis gas. Sandia Report SAND. 2007. April 2007

[9] Yu YH. Simulation of secondary reformer in industrial ammonia plant. Chemical Engineering and Technology. 2002;**25**:307-314

[10] Ib D. New reforming concepts for large scale NH_3 plants. Haldor Topsøe A/S Ammonia Technical Manual. 2005: 238-254

[11] Gardner DC, Bartholomew CH. Kinetics of carbon deposition during methanation of carbon monoxide. Industrial and Engineering Chemistry Product Research and Development. 1981;**20**:80-87

[12] Wright RE, Wolff HI. Refractory problems in production of hydrogen by pyrolysis of natural gas. Journal of the American Ceramic Society. 1948;**31**: 31-38

[13] Lim JY, McGregor J, Sederman AJ, Dennis JS. The role of the Boudouard and water gas shift reactions in the mechanism of CO or CO_2 over Ni - γ-Al_2O_3 catalyst. Chemical Engineering Science. 2016;**141**:28-45

[14] Strait Megan, Allum Glenda, Gidwani Nisha. Synthesis Gas Reformers. Available from: https://www.owlnet.rice.edu/~ceng403/nh3ref97.html

[15] Tso ST. The Corrosion of Silicate Materials by Hydrogen Gas and Hydrofluoric Acid Solution [PhD thesis]. USA: University of California Berkeley; 1979

[16] Crowley MS. Hydrogen-silica reactions in refractories-part II. Bulletin of the American Ceramic Society. 1970; **49**:527-530

[17] Ming-Chu C, Cutler IB. Vaporization of silica in steam atmosphere. Journal of the American Ceramic Society. 1979;**62**: 593-596

[18] Crowley MS. Hydrogen-silica reactions in refractories. Bulletin of the American Ceramic Society. 1967;**48**: 679-682

[19] Palmer G. Volatilization of refractory silica in hydrogen water vapour gas streams. Refractories Worldforum. 2012;**4**:63-70

[20] Robbins CR, Mauer F. A, chemical degradation of castable refractories in coal gasification process environments. Journal of Materials for Energy Systems. 1981;**3**:32-42

[21] Lindstad T, Syvertsen M, Ishak RJ, Arntzen HB, Grøntvedt PO. The influence of alkalis on the Boudouard reaction. In: Proceedings: Tenth International Ferroalloys Congress (INFACON X); 1–4 February 2004; Cape Town, South Africa. pp. 261-271

[22] Opila EJ. Oxidation and volatilization of silica formers in water vapor. Journal of the American Ceramic Society. 2003;**86**:1238-1248

[23] Kennedy CR, Schlett PE. Refractories for coal gasification. The state-of-the-art in the U.S. In: Kröckel H, Merz M, Van Der Biest O, editors. Ceramics in Advanced Energy Technologies. Dordrecht: Springer; 1984. pp. 283-317. DOI: 10.1007/978-94-009-6424-2_16

[24] Paul XM-W, Brown JJ. Mechanism of iron catalysis of carbon monoxide decomposition in refractories. Journal of the American Ceramic Society. 1989;**72**: 110-115

[25] Davis Robert F, Aksay Ilhan A, Pask Joseph A. Decomposition of Mullite. Journal of the American Ceramic Society. 1972;**55**:98-101

[26] Poirier J, Kadok J, Bost N, Coulon A, Ammar MR, Brassamin S, et al. Inhibiting the carbon deposition from the reverse Boudouard reaction in refractories submitted to CO-H_2 atmosphere. Refractories Worldforum. 2019;**11**:83-91

[27] Day Delbert E, Gac FD. Stability of refractory Castables in steam - N_2 and steam-CO atmospheres at 199°C. Bulletin of the American Ceramic Society. 1977;**56**:644-648

[28] Walker PL, Rakszawski JF, Imperial GR. Carbon formation from carbon monoxide-hydrogen mixtures over iron catalysts-II. Rates of carbon formation. The Journal of Physical Chemistry. 1959;**63**:140-149

[29] Ellingham HJT. Reducibility of oxides and sulphides in metallurgical processes. Journal of the Society of Chemical Industry, London. 1944;**63**: 125-160. DOI: 10.1002/jctb.5000630501

Section 2

Gasification Processes

Chapter 7

Recent Advances in Supercritical Water Gasification of Pulping Black Liquor for Hydrogen Production

Changqing Cao, Lihui Yu, Wenhao Li, Lanjun Liu and Peigao Duan

Abstract

Black liquor is the wastewater produced from paper-making-pulping industry, which has great threats to the environment and human health. Its pollution handling and resourcing attracted much attention, but conventional method showed several drawbacks, including low energy efficiency, emission of secondary pollutants, and safety and operating issues. Supercritical water gasification (SCWG) is a promising method for organic wastes, especially for those with high moisture content, so it has been widely investigated. A previous study showed that SCWG of pulping wastewater can realize pollution elimination and hydrogen production simultaneously. In this chapter, the recent development of this technology in past decade will be reviewed, including the gasification performance, the influence of the main operating parameter, the catalysts used in this process, the synergetic effect in co-gasification with other energy sources, and the evaluation of integrated system of SCWG of black liquor with pulping process. These progresses have been made will boost the industrial utilization of SCWG of black liquor in pulping industry.

Keywords: black liquor, supercritical water gasification, hydrogen, catalysts, co-gasification

1. Introduction

Hydrogen is a clean energy carrier and favorable in relieving pollution, which gained much attention. Hydrogen production from biomass and organic wastes has great potential for sustainable development. Black liquor is the wastewater from the paper-making pulp industry and mainly contains inorganic pulping chemicals (NaOH and Na_2CO_3), organic components (lignin, hemicellulose, and their derivatives), and water [1]. It has the properties of high alkalinity, high chemical oxygen demand (COD) content, offensive odor, dark caramel color, and high moisture, so it will bring great threat to the environment and human health if not treated properly. Approximately 170 million tons of black liquor solids are generated annually [2]. Traditionally, black liquor is treated by combustion or conventional gasification [3, 4]. Both methods require an energy-intensive evaporation process, which leads to the waste of energy and production of secondary pollutants, such as NO_x, SO_2, and fine particles. Alternatively, the electrochemical treatment of black liquor is a clean method, but it requires high consumption of valuable electric power [5]. Therefore, more efficient and cleaner technology is desired as an alternative for handling black liquor.

Supercritical water gasification (SCWG) is a promising technology that can convert biomass and wastewater into hydrogen-rich gas. Relying on its unique physicochemical properties, SCW provides an excellent reaction environment for the gasification of biomass and organic wastes. Compared with traditional handling methods of black liquor, this technology has several advantages. For example, it reduces the energy dissipation of evaporation and allows wastewater with high moisture as feedstock, the alkali present in the black liquor can be served as a catalyst, no pollutants are generated, and the products are easy to separate and purify. Therefore, it has attracted the attention of many researchers and much progress has been made in this decade, and many studies have been published on this topic.

In this chapter, recent advances in SCWG of black liquor in recent decades will be reviewed. First, the effect of operating parameters, including temperature, concentration, and wall material, is discussed. In addition, the recovery of alkali salts in black liquor is introduced. Second, the catalysts used in this process and synergetic effect in its co-gasification with coal, biomass, and plastics are presented. Third, the poly-generation system of SCWG of black liquor is evaluated.

2. Gasification performance of black liquor

2.1 Influence of reaction temperature

Temperature is a key parameter affecting the gas product composition and gasification efficiency in SCWG of black liquor. From the thermodynamic aspects, high temperatures tend to increase gasification efficiency and hydrogen production, while methane dominates the gas products at lower temperatures [6]. We performed thermodynamic analysis of SCWG of black liquor at different temperatures (300–800°C) with a pressure of 25 MPa and found that the temperature can greatly affect the gas composition [7]. CH_4 and CO_2 were the main components at low reaction temperatures, but their content decreased with increasing temperature. The fraction of H_2 increased with temperature, which was above 50% at temperatures over 650°C, and the maximum equilibrium H_2 fraction of 61.34% was obtained at 800°C. Gasification efficiency higher than 100% was attributed to the participation of water, which was also increased with temperature. The decrease in HHV (higher heating value) of the gas product with increasing temperature was because the fraction of H_2 (HHV = 12.75 MJ/Nm^3) increased, while the fraction of CH_4 (HHV = 39.82 MJ/Nm^3) decreased.

The experimental study also showed that temperature is a critical factor on the gas composition and yield. Boucard et al. [8] studied the gasification of 10 wt% black liquor at different temperatures (350–450°C, 25 MPa, and 60 min) and found the C content in the gas phase increased from less than 1% at 350°C to more than 25% at 450°C. The H_2 yield increased from 2 mol/kg under subcritical conditions to 10–40 mol/kg under supercritical conditions. Previously, we studied SCWG of 9.5 wt% wheat straw black liquor at 400–600°C and 5 kg/h in a continuous flow system [9]. When the temperature was increased from 400°C to 600°C, the gas yield nearly doubled, and gasification efficiency and H_2 yield increased from 28.05% and 6.82 mol/kg to 67.89% and 11.26 mol/kg, respectively. The COD concentration and pH decreased from 95,000 mg/L and 11.3 in the raw black liquor to 2160 mg/L and 7.0–7.8 at 600°C, respectively. In addition, the color also changed from dark caramel to clear as pure water (**Figure 1**). SCWG of black liquor at higher temperatures (600–750°C) was investigated in a batch reactor [7]. The H_2 fraction increased from 55% to nearly 70% as the temperature increased from 600°C to 750°C. And the maximum carbon gasification efficiency of 94.10% was obtained at 750°C and 30 min.

2.2 Influence of black liquor concentration

The concentration of weak black liquor produced from pulping varied in the range of 10–20 wt% [10], which can influence the gasification performance. **Figure 2** showed the influence of black liquor concentration on the equilibrium composition, yield, and HHV of the gas product. The effect of concentration on gas composition was almost opposite to temperature. Increasing the concentration greatly decreased the H_2 fraction but increased the fraction of carbonaceous gas (CO_2, CH_4, and CO), so increasing the concentration increased the heating value of the gas product.

In the experiments with both continuous systems and batch reactors, high H_2 yields and fractions were obtained from gasification of diluted black liquor. The gas yield

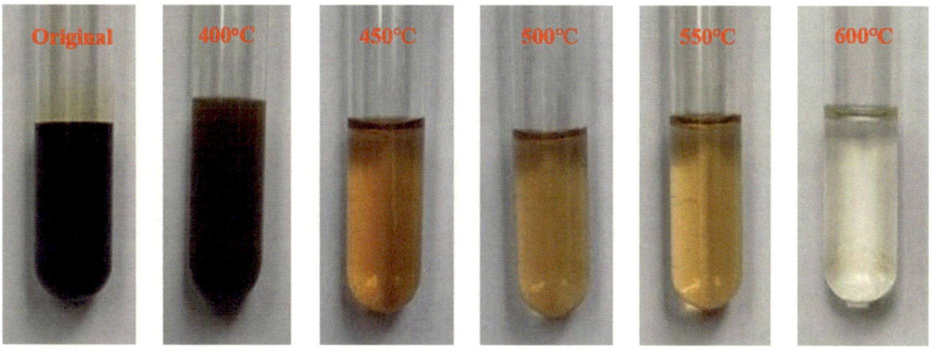

Figure 1.
The color changes of black liquor after SCWG in a tubular reactor at different temperatures (concentration = 9.5 wt%; pressure = 25 MPa) [9].

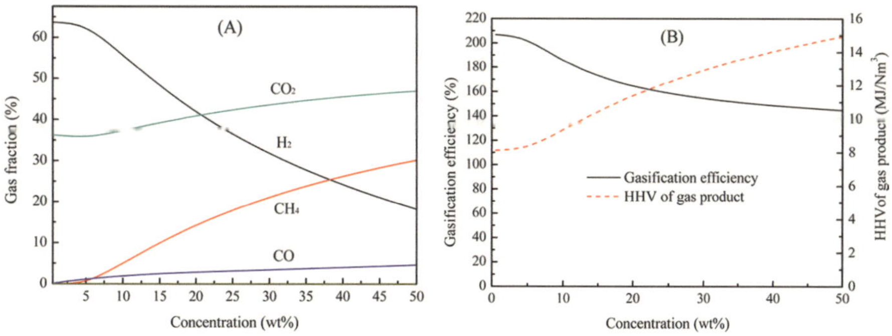

Figure 2.
Equilibrium gas composition (A) and GE and HHV of the gas product (B) of wheat straw black liquor with change of concentration (pressure = 25 MPa; concentration = 9.5 wt %) [7].

almost doubled when the concentration decreased from 9.5 wt% to 3 wt% at 550°C [7, 11]. Casademont et al. also found that the gas yield and H_2 yield increased more than fourfold as the black liquor concentration decreased from 2.43 wt% to 0.81 wt% in SCWG at 600°C [12]. At higher concentrations, reducing the concentration from 20 wt% to 10 wt% increased hydrogen in black liquor converted to H_2 from 7.5 to 9.8% at 650°C [1]. Moreover, the H_2 yield obtained by gasification of low-concentration black liquor (below 3 wt%) can reach over 24 mol/kg, while the gas products from gasification of above 10 wt% black liquor were mainly composed of C_1–C_4 hydrocarbons [1, 9, 12]. Therefore, high concentration black liquor is more difficult to gasify. However, a lower concentration of black liquor will improve the system scale and energy loss, so a proper concentration needs to be selected through further detailed investigation.

2.3 Influence of reactor wall material

In SCWG, water will fill the entire space of the reactor and Ni was an effective catalyst in SCWG [13], so the material of the reactor wall can also affect the gasification performance. Casademont et al. [12] obtained different gasification efficiencies of black liquor in SCWG from the literatures [7, 9], and they attributed the difference mainly to the different reactor materials. They proposed that the reactor made of Inconel 625 was more favorable to gasification than other material. De Blasio et al. [14] investigated SCWG of black liquor in stainless steel 316 and Inconel 625 reactors. It was found that the carbon gasification efficiency obtained was similar in these two reactors, while the hydrogen gasification efficiency obtained in Inconel 625 was much higher than that of the stainless steel 316 reactor at both 600 and 700°C. Hydrogen production with the Inconel 625 reactor was also much higher than that with stainless steel 316 reactor at 600°C. Inconel 625 was slightly better than stainless steel 316 in hydrogen production at 500°C and close to each other at 700°C. Besides, the hot gas efficiency obtained with Inconel 625 reactor was higher than that with stainless steel 316 at 600–700°C, and the maximum value exceeded 80% at 700°C. In a word, the treatment of black liquor in the Inconel 625 reactor can increase hydrogen production and inhibit the formation of tar and char. Özdenkçi et al. [15] investigated the technoeconomic feasibility of SCWG of black liquor in Inconel 625 and stainless steel 316 reactors. It showed that Inconel 625 outperformed stainless steel 316 in terms of energy production, hydrogen production, resistance to pulping chemical losses, and changes in energy prices.

2.4 Influence of alkali in black liquor

The role of alkali contained in black liquor in the gasification process was investigated by comparing the effect of black liquor and lignin as additives on SCWG of coal [16]. It was found that the fraction of CO decreased significantly, while the content of H_2 and CO_2 increased when the additive was changed from lignin to black liquor. It was probably because the alkali in black liquor promoted the water-gas shift reaction. The total gas yield of coal with black liquor (12.07 mol/kg) was higher than that of coal with lignin (6.90 mol/kg). Hawangchu et al. [17] investigated the effect of inherited alkali on product distribution by comparing the gasification performance of soda black liquor, kraft black liquor, and lignin compound in SCW. Hydrogen from soda black liquor was always higher than that of lignin, confirming the promotion of alkali on the water-gas shift reaction. They proposed that both the dissolution of organic substances and inhibition of coke production by alkalis promoted gas production.

In addition, Rönnlund et al. [18] studied the effect of the addition of alkalis (KOH, K_2CO_3, NaOH) and black liquor on SCWG of paper sludge. Similarly, the addition of black liquor improved the hydrogen production and gasification efficiency of paper sludge gasification. Furthermore, the promotion extent of black liquor on gasification was similar to that of alkali salts.

3. Recovery of alkali in SCWG of black liquor

Alkali salts are another main component of black liquor, including cooking chemicals (NaOH and Na_2S), as well as derived salts such as Na_2CO_3, Na_2SO_4, and Na_2SO_4 [19]. The organic components in black liquor can be converted in SCWG into gas products, while the inorganic salts still reside in the reactor. Thus, the salts can be recovered and reused in the pulping process as cooking chemicals to reduce the pulp cost. In conventional treatment, the recovery of alkali can be realized by burning the organic components in black liquor [20]. Depending on the unique physicochemical properties of SCW, the alkali can also be recovered during SCWG. The literature showed that the dielectric constant of water decreases sharply with increasing temperature around the critical point [21], which reduces the solubility of inorganic substances and enables the recovery of alkali salts.

Alkali was the main cooking chemical (NaOH), so the recovery of Na^+ salts in SCWG of black liquor was investigated [22, 23]. We studied the distribution of alkali during SCWG in a fluidized-bed reactor using glucose as the model compound and attempted to recover the alkali salts [22]. For the extremely low solubility, the Na^+ salts were mainly precipitated in SCW and distributed in the reactor during gasification, which were not carried out of the reactor by the fluid in the form of ions [22]. While the fluid temperature in the reactor was reduced close to the critical point after the gasification, the alkali was dissolved in the water and flowed out as the reaction effluent. As a result, the Na^+ content and pH value of the effluent changed dramatically with the temperature in the cooling process of the reactor (**Figure 3**). The Na^+ content of the effluent was 10–30 mg/L when the temperature was higher than 360°C, and it increased sharply when the temperature dropped to 355°C and reached a maximum value of 1815 mg/L at 335°C. With flushing of the reactor for a certain time, the Na^+ content dropped to 86 mg/L. Cooling fluid in the range of 360–200°C achieved

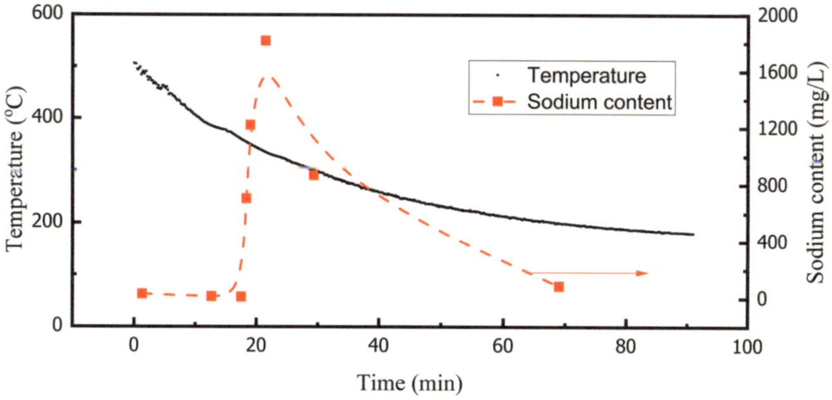

Figure 3.
The Na^+ content of effluent varies with the temperature of the fluid [22].

a Na^+ recovery of 81.07%, and the losses may be attributed to salt entrainment. Based on this result, we proposed an alkali recovery method during SCWG of black liquor, in which the alkalis can be recovered by cooling the reactor after operating for a certain time. And for the fluidized-bed reactor, the velocity of the fluid was important to constrain the alkali solids in the reactor and reduce the alkali salt entrainment and losses. Besides NaOH, the inorganic components of the effluent also contained Na_2CO_3 and $NaHCO_3$ generated by the reaction of NaOH with CO_2. Regeneration from $Na_2CO_3/NaHCO_3$ to NaOH can be achieved by a conventional method (causticization reaction) with CaO [22]. Some methods can be used to improve the recovery of NaOH in SCWG of black liquor. Fedyaeva et al. [23] studied SCWG of Na_2CO_3-free black liquor and calculated the mole ratio of $2NaOH/Na_2CO_3$. The ratio could reach 4 mol% without water flow. The amount of produced Na_2CO_3 can be controlled by pumping SCW through the reactor to carry out CO_2 and reduce the residence time of SCW. The ratio can reach 14–18% at 710–750°C and 30 MPa. It is a simple and effective method to achieve highly economic recovery of alkali salts without the addition of chemicals.

4. Catalysts used in SCWG of black liquor

To improve the gasification efficiency and lower the reaction temperature, the addition of catalysts can be considered. Catalysts for SCWG can be classified into alkali salts, metals, and metal oxides [24]. They have different advantages in SCWG of biomass due to their different catalytic properties. In particular, alkali salts are effective catalysts for SCWG of biomass [25]. However, some catalysts, including alkali salts, may be inapplicable in SCWG of black liquor due to its unique composition and properties (**Table 1**).

4.1 Alkali

Alkalis are widely used for enhancing hydrogen production in SCWG due to their strong promotion of the water-gas shift reaction and C–C bond cleavage [24, 26, 27]. With NaOH, the fraction of H_2 can reach over 50%, and the H_2 volume increased to 80% in SCWG of glucose [27]. Compared with noncatalytic gasification, the H_2 yield from SCWG of sugarcane bagasse can be increased to over 10 times with the catalysis of five alkali salts [28]. The highest H_2 yield of 75.6 mol/kg was obtained under the catalysis of KOH at 800°C. Comparatively, the catalytic effect of alkali salts on SCWG of black liquor is weaker. Casademont et al. [12] studied SCWG of black liquor (0.81 wt%, pH = 11.12) in a continuous tubular reactor with NaOH as the catalyst.

	Advantages	Disadvantages
Alkali	Homogeneous, strong promotion on water-gas shift reaction and C–C bond cleavage	Hard to recycle and regenerate, weak catalytic effect
Metal	High catalytic activity, strong selectivity, thermal stability	Expensive, deactivation, hard to regenerate
Metal oxide	Stable, low-cost, recyclable and regenerate, unlimited loading	Less active, hydrogen consumption

Table 1.
Comparison of different types of catalysts in SCWG of black liquor.

The gasification efficiency slightly improved from 36.5% to 37.0% with the addition of 0.6 wt% NaOH, but the H_2 yield decreased from 24.58 to 23.34 mol/kg at 600°C. Moreover, when the NaOH loading increased to 1.2 wt%, the H_2 yield decreased to 22.27 mol/kg, while the gasification efficiency increased slightly to 44.3%. When the temperature was raised to 700°C, the H_2 yield increased to a maximum of 38.68 mol/kg with 0.9 wt% NaOH from 33.62 mol/kg without catalyst. They attributed the weak effect of alkali to the abundant alkali contained in black liquor, which made the addition of alkali ineffective. Guo et al. [29] also studied the effect of NaOH on SCWG of glycerol in a continuous reactor and found that the gasification efficiency increased slightly with over 0.1 wt% alkali. They proposed that excess alkali will not dissolve in SCW to improve the catalytic effect but will accelerate plugging due to salt deposition. In addition, the absorption of CO_2 by OH^- increased the chemicals consumption or regeneration time from Na_2CO_3 to NaOH. The addition of potassium will lower the melting point of salts and cause operating and safety problems [30]. Therefore, alkali was not a suitable catalyst for SCWG of black liquor for the high content of alkalis in itself.

4.2 Metals

Some metals, such as Pb, Pt, Rh, Ni, and Ru, are effective catalysts in SCWG of biomass and organic wastes [31, 32]. However, they were prone to be deactivated with the presence of sulfur for poisoning. Peng et al. [33] studied the performance of Ru/C catalyst in SCWG of sulfur-containing (0.53 wt%) microalgal and found that the BET surface of the bottom catalyst decreased from 1254 m^2/g to 56 m^2/g after 55 h. The activity of the catalyst was severely reduced. The other notable change was that the amount of S contained in the catalyst significantly increased from 151 mg/kg to 1156 mg/kg. They proposed that sulfur poisoning was the main cause of catalyst deactivation. Osada et al. [34, 35] investigated SCWG of lignin catalyzed by Ru, Rh, Pt, and Pd in the presence of sulfur and found deactivation of the catalysts due to sulfur poisoning. It was found that sulfur will combine with Ru and chemically adsorb on the catalyst surface, which will decrease the active sites and reduce the gas yield. Black liquor obtained from Kraft pulping contains a high fraction of NaOH and Na_2S [1, 23, 36], which will deactivate the metal catalyst and increase the operating cost.

Appropriate supports can improve the stability of metal catalysts in harsh environments and increase the dispersion of active components, such as using oxides, zeolites, and carbons as the support [37]. However, the supports may be unstable in SCW. For example, γ-Al_2O_3 was converted to α-Al_2O_3 in a hydrothermal environment, and zeolites undergo structural collapse in SCW [38, 39]. The phase transformation and structural collapse of the support can reduce catalytic activity.

4.3 Metal oxides

Compared with alkali and metal catalysts, metal oxides show unique advantages and can be used in SCWG of black liquor. Metal oxides are easier to recover than alkali salts, and they are more stable and lower-cost than metals. It allows metal oxides to be stored and transported for a long time. Unlike metal catalysts, metal oxides can act as sulfur fixers and exhibit excellent activity [33, 40]. Ates et al. [41] studied the role of ZnO and MoO_3 in SCWG of model feed (dibenzothiophene-hexadecane). The conversion and sulfur removal of dibenzothiophene greatly increased with the metal oxides. The catalysts used consisted of metal oxides and sulfided metal oxides.

Moreover, sulfur-fixing metal oxide can be regenerated by a facile calcination treatment, whereas metal catalysts cannot [42].

Some metal oxide exhibit high catalytic activity in SCWG of biomass. Park et al. [43] studied RuO_2-catalyzed SCWG of organic components and achieved nearly complete gasification of organic compounds. They proposed that the redox cycle between Ru^{IV} and Ru^{II} was the mechanism of the catalytic impact. Especially, some transition metal oxides can catalyze water-gas shift reaction to improve hydrogen production. The gasification efficiency of glucose and cellulose under the catalysis of ZrO_2 was found to be approximately twice that obtained under noncatalytic condition [44]. Seif et al. [45] used Co_3O_4, CuO, and MnO_2 to catalyze SCWG of industrial wastewaters for hydrogen production in a batch reactor. The catalysts increased the hydrogen yield by promoting water-gas shift reaction. The largest improvement in H_2 yield was from 0.46 mol/kg without catalyst to 1.25 mol/kg with 40 wt% Co_3O_4 catalyst at 350°C. The order for increasing hydrogen production was $Co_3O_4 > CuO > MnO_2$. In our previous study, 12 transition metal oxides were screened for SCWG (600°C, 25 MPa, and 10 min) of 10 wt% glucose for hydrogen production in quartz capillary reactors [46]. Most of the metal oxides improved gasification efficiency. The highest carbon gasification efficiency of 87.11% was obtained by Cr_2O_3 compared with that without catalyst (68.27%). The highest H_2/CO ratio up to 19.61 and 14.55 was obtained with the promotion of CuO and MnO_2 on water-gas shift reaction.

The application of metal oxide catalysts in SCWG of black liquor has also been studied. Boucard et al. [8] used nano-CeO_2 to enhance the conversion of 10 wt% black liquor in subcritical (350°C and 25 MPa) and supercritical (450°C and 25 MPa) water in a batch reactor. The composition of the gas product was H_2, CO, CO_2, and light hydrocarbons. The catalyst improved gasification efficiency and hydrogen production. The H_2 yield increased from 322 to 345 μmol under supercritical conditions. CeO_2 promoted the splitting of SCW into hydrogen and oxygen, where the hydrogens combined with each other to produce H_2, and complete oxidation could be achieved with oxygen. Then, the degradation of black liquor was promoted, and the color changed from dark brown to yellow and was almost transparent. Under subcritical condition, the H_2 yield increased by 10 times with the catalyst but was lower than that under supercritical condition. The catalyst promoted water splitting with lower efficiency under this condition, so a low H_2 yield and partial oxidation were achieved [8].

In our previous study [11], hydrogen production by SCWG of 12.6 wt% black liquor with 14 common metal oxides (ZnO, TiO_2, SnO_2, V_2O_5, WO_3, MoO_3, Fe_3O_4, Fe_2O_3, MnO_2, Cr_2O_3, CeO_2, CuO, ZrO_2, and Co_2O_3) was studied at 600°C. The gasification efficiency of black liquor was 77.7% without catalyst, and the gasification efficiency was increased to over 100% with over half of these catalysts. As the gasification efficiency was defined as the ratio of the mass of the gas product to the mass of black liquor solids, it indicated that the presence of these catalysts promoted the involving of water in the reaction, which contributed more in hydrogen production. The gasification efficiency obtained with WO_3, Co_2O_3, and V_2O_5 was 111.84%, 111.82%, and 110.06%, respectively. All the catalysts improved the carbon gasification efficiency, among which WO_3, Co_2O_3, and V_2O_5 exhibited the highest activity, and the carbon gasification efficiency increased from 50.85% without catalyst to 71.05%, 70.92%, and 67.98%, respectively. Most metal oxides also increased hydrogen production by promoting the water-gas shift reaction. The presence of SnO_2, ZnO, and Co_2O_3 increased the H_2 fraction from 30.76% to 56.79%, 54.9%, and 54.8%, respectively.

The highest H_2 yields of 21.67 mol/kg and 21.03 mol/kg were obtained with Co_2O_3 and ZnO, respectively. From the characterization of the metal oxides before and after being used in SCWG, the authors proposed that catalysts may provide oxygen for reactions and promote degradation of black liquor (**Figure 4**). The high activity of V_2O_5 in improving gasification efficiency may be due to its highest oxygen content, which also led to the highest CO_2 production. The activity of Fe_2O_3 was higher than that of Fe_3O_4, and the yield of CO_2 was higher. TiO_2, as the catalyst with the oxygen content second to V_2O_5, also increased the CO_2 fraction. The effect of catalyst loading on gasification was also investigated. Increasing the amount of catalyst from 0 wt% to 180 wt%, the gasification efficiency continued to increase without an upper limit. It seems to be a unique advantage of metal oxides compared with alkali salts and metal catalysts [47–49]. The maximum carbon gasification efficiency (80.11%) and gasification efficiency (149.90%) increased by nearly 1.6 times and two times, respectively. Sufficient metal oxides reduced the H_2 fraction and increased the CO_2 fraction. This may be related to the catalytic mechanism of metal oxides in SCW. Metal oxides can supply oxygen species for reactions and enhance the degradation of black liquor. Excessive oxygen may consume hydrogen and reduce H_2 production. Residual low-valence metal oxides or metals still have catalytic activity, which can promote water-gas shift reaction to improve gasification and H_2 production.

The synergistic and complementary catalysis brought by bimetallic oxides is remarkable [50, 51]. It can also play a well catalytic role in SCWG of biomass. Mastuli et al. [52] studied SCWG of oil palm frond biomass with the catalysis of NiO, CuO, and ZnO supported on MgO and found that the catalysts increased gas yield from

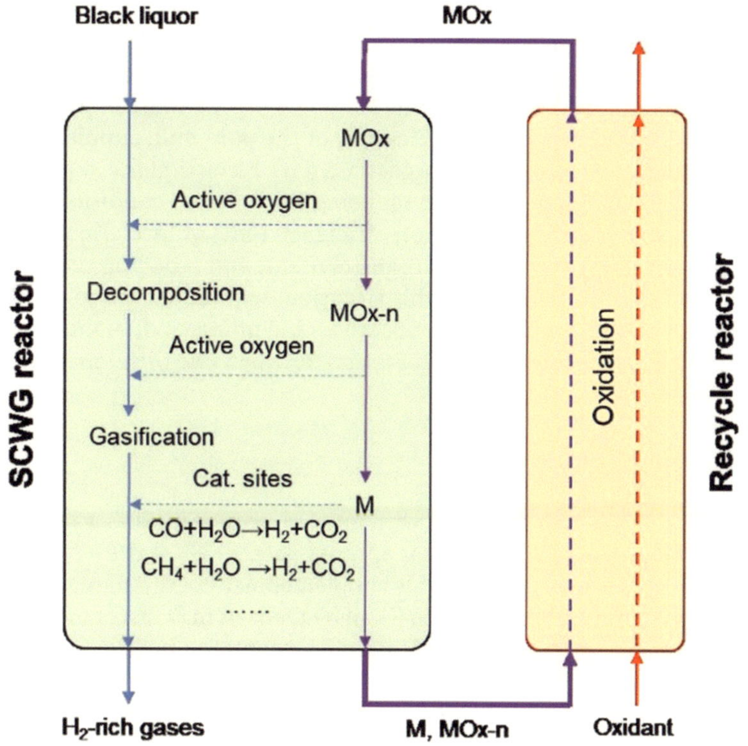

Figure 4.
Catalytic mechanism of metal oxides in SCWG of black liquor [11].

55.4 mol/L with MgO to 72.7, 81.1, and 118.1 mol/L under the addition of NiO/MgO, CuO/MgO, and ZnO/MgO, respectively. The H_2 yield was improved from 35.4 mol/L without catalyst to a maximum of 118.1 mol/L under the promotion of ZnO/MgO. They proposed that Zn-based catalyst was more effective than Ni- and Cu-based catalysts in promoting water-gas shift reaction. The catalytic effect of bimetallic oxides on lignin, one of the organic components of black liquor, was also studied. We used CeO_2-ZrO_2 as catalyst in SCWG of lignin and cellulose at 500°C and 600°C [53]. The gasification efficiency of lignin increased from 57.71% to 70.43% with catalyst at 600°C and the highest H_2 yield increased to 20.39 mol/kg. The catalytic effect of CeO_2-ZrO_2 at a lower temperature was more significant. This may be because at lower temperatures, the gasification intermediate products were mainly distributed in the liquid, which can be fully contacted with the catalyst. In SCWG of cellulose, the catalyst also increased gasification efficiency and hydrogen yield. The H_2 yield increased nearly 2.6 times from 3.29 mol/kg without catalyst to 8.5 mol/kg at 500°C and increased to 19.39 mol/kg at 600°C. Similarly, the catalysis of CuO-ZnO and Fe_2O_3-Cr_2O_3 in SCWG of lignin and cellulose was investigated [54]. In SCWG of lignin by CuO-ZnO, the highest gasification efficiency of 96.24% and the highest H_2 yield of 29.62 mol/kg was obtained at 600°C. The order of catalysts in enhancing hydrogen production from SCWG of lignin is CuO-ZnO > Fe_2O_3-Cr_2O_3 > CeO_2-ZrO_2. In the gasification of cellulose, the catalytic effect of Fe_2O_3-Cr_2O_3 on H_2 yield was more significant than that of CuO-ZnO, reaching a maximum of 18.74 mol/kg.

5. Co-gasification of black liquor with other energy substances

SCWG is a promising technology to convert black liquor into hydrogen-rich gas, but our previous study [55, 56] showed that the energy produced from SCWG of black liquor cannot fully meet the requirement of the pulp mill. To produce more energy and realize the energy self-sufficiency of the pulp mill, supplementary energy resources can be mixed and co-gasified with black liquor. Co-gasification also has a series of advantages, such as reducing the reaction conditions by synergistic effects, improving the gasification efficiency (GE) of black liquor and other energy substances, increasing the hydrogen content, and reducing the cost of hydrogen production. In response to this situation, we studied the co-gasification of black liquor with different energy substances including coal, biomass, and waste plastics (**Figure 5**) and studied the interaction between two different materials in co-gasification.

5.1 Co-gasification with coal

Coal is an important energy source with huge reserves in China, with approximately 162.29 billion tons accounting for approximately 13% of the world's total reserves [57]. Coal can be used as a stable supplementary energy source for pulp mills, and clean and efficient conversion of coal can be realized in SCWG at the same time [58, 59]. Inherent alkali in black liquor is an efficient catalyst in SCWG of the organics that can suppress the formation of tar and coke and increase hydrogen production, so co-gasification of black liquor and coal can utilize the inherent alkali in black liquor to enhance coal gasification.

Co-gasification of coal and black liquor in SCW was studied by thermodynamic analysis and experiments [16]. Thermodynamic analysis showed that the mixing

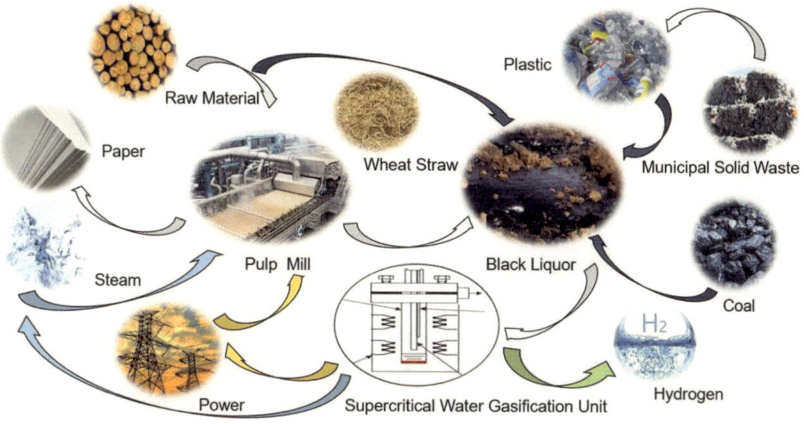

Figure 5.
Schematic diagram of co-gasification of black liquor and other energy substances.

of coal with black liquor can enhance H_2 production in the equilibrium state. The experiments performed in a fluidized-bed reactor at 550°C and 25 MPa showed that the presence of coal and the ratio of coal/black liquor influenced the gasification (**Figure 6**). With increasing black liquor fraction, the H_2 fraction and GE increased, while the CO fraction decreased. When the black liquor fraction was increased from 0 to 50 wt%, GE increased from 15% to above 60%; the H_2 fraction increased from 34 to 45%; and the CO fraction decreased from 20% to below 1%. We proposed that the improvement in the GE of coal with the addition of black liquor was due to the alkali and lignin contained in black liquor. The alkali in black liquor not only catalyzed the decomposition of coal but also accelerated the water-gas shift reaction [59], so the gasification efficiency and hydrogen product were improved with increasing black liquor fraction. On the other hand, the lignin in black liquor may also promote coal gasification during SCWG. Lignin will decompose before coal due to unique components and molecular structure, so the generated phenolic compounds from lignin will facilitate the extraction of organics and decomposition of coal [60].

We also investigated the co-gasification performance of coal and black liquor at high temperatures (600–750°C) [61]. The variation in the temperature greatly changed the gasification efficiency. The CE and GE increased from 36.88% and less

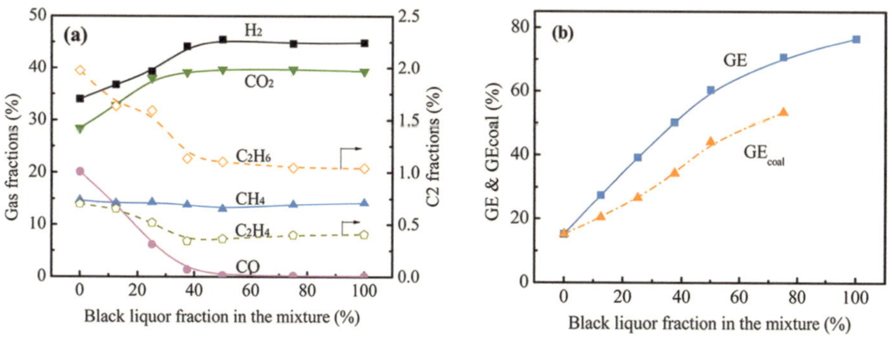

Figure 6.
Experimental results of SCWG of coal-black liquor mixtures with different black liquor fractions: (a) gas fractions; (b) GE (T = 550°C; P = 25 MPa; total concentration = 10 wt%) [16].

than 80% to 79.46% and over 180%, respectively. The highest CE (79.46%) was obtained for the co-gasification reaction of coal with black liquor with a mixing ratio of 50:50 at 750°C. With increasing temperature, the PAHs can be further decomposed, thereby improving CE and GE. In addition to increasing temperature, the use of fluidized-bed reactors and catalysts can also improve GE at lower temperatures. Therefore, higher gasification efficiency can be obtained with the combination of the proper reaction conditions, catalyst, and advanced reactor.

5.2 Co-gasification with biomass

As biomass is the raw material for pulp production, which is readily available in pulp mills and is also inexpensive and renewable. Therefore, biomass was considered as an available supplementary energy source to co-gasify with black liquor to provide more energy. As wheat straw is an important pulping raw material in China, we studied the co-gasification of black liquor with wheat straw in a batch reactor at 500–750°C with different mixing ratios [62]. The addition of black liquor and the change in the mixing ratio significantly influenced the gasification performance. As the black liquor fraction increased from 25 wt% to 100 wt%, both the H_2 yield and CO_2 yield decreased by approximately 7 mol/kg. The CE and GE of co-gasification were higher than the theoretical value calculated from the separate gasification of black liquor and wheat straw, indicating that a synergistic effect existed in co-gasification. The synergistic effect was probably because the inherent alkali in black liquor had a catalytic effect on wheat straw, and the presence of wheat straw can utilize the alkalis to the maximum extent. When the mixing ratio was 50:50, the highest CE and GE of close to 90% and 140% were obtained, respectively, indicating that it is the optimal mixing ratio for co-gasification under this reaction condition.

Comparatively, the synergistic effect of black liquor with wheat straw [62] was smaller than that of black liquor with coal [16, 61]. At a mixing ratio of 50:50, the GE and H_2 yields of co-gasification of black liquor and wheat straw were approximately 30% and 20% higher than those in co-gasification of black liquor and coal. The difference was mainly attributed to the different compositions and properties of the organic matter. Compared with coal, wheat straw was more similar to black liquor because it was the source of black liquor. The larger difference between black liquor and coal may lead to the generation of different reaction intermediates and environments, which is more beneficial for the interaction of their gasification and synergistic effect.

5.3 Co-gasification with waste plastics

Besides, the municipal solid wastes (MSW) generated from nearby communities were also a good option for the energy supplement. In MSW, plastic waste is an important component, and its treatment is one of the important issues [63]. Co-gasification of plastic wastes and black liquor not only improves the energy production from SCWG of black liquor but also achieves efficient and clean conversion of waste plastics into hydrogen-rich gas. Hence, we studied the co-gasification of different plastic wastes (PE, PC, PP, ABS) and alkali lignin, which is the main component of black liquor.

The co-gasification of alkali lignin with various plastics (PE, PC, PP, and ABS) in SCW was investigated in a batch reactor [64]. The gasification efficiencies of the four plastics were different due to their different degrees of polymerization and chemical structures, and the order of magnitude was PE > PC > PP > ABS. In the separate gasification, the CE and GE (58.59% and 109.90%) of alkali lignin were higher than those of polyethylene (53.27% and 91.38%), but both were lower than the values obtained in the co-gasification. It indicated that there was a synergistic effect in co-gasification of alkali lignin with polyethylene. This was also supposed to be related to their different chemical structures and properties of different organic substances (**Figure 7**). During co-gasification, alkali lignin can promote the depolymerization of polyethylene into different fragments, and the generated intermediates will further be decomposed into large amounts of CH_4, C_2H_4, and C_2H_6. Alkali lignin can also catalyze steam reforming of these gaseous products to H_2 and CO_2. In addition, polyethylene had a higher H/C ratio than lignin, which can provide more hydrogen radicals and inhibit the cross-linking and repolymerization of aromatic compounds during lignin decomposition. Therefore, co-gasification of alkali lignin with polyethylene can not only inhibit the formation of char, but also promote the gasification of each other.

6. System analysis on SCWG of black liquor

Compared with conventional treatment methods, the hot and compressed products from SCWG of black liquor include combustible gases such as H_2, CH_4, and CO and contain considerable chemical energy and sensible heat, which was necessary to be recovered to improve the energy efficiency. On the other hand, pulping is a complex process involving multiple steps, including raw material preparation, pulping, bleaching, chemical recovery, and pulp drying, which consume various types of energy, including power, medium-pressure (MP) and low-pressure (LP) steam [55]. Therefore, it is necessary to integrate SCWG of black liquor and the pulping process to make full use of the energy and improve the energy efficiency, which can also provide a reference for process designing for its industrialization.

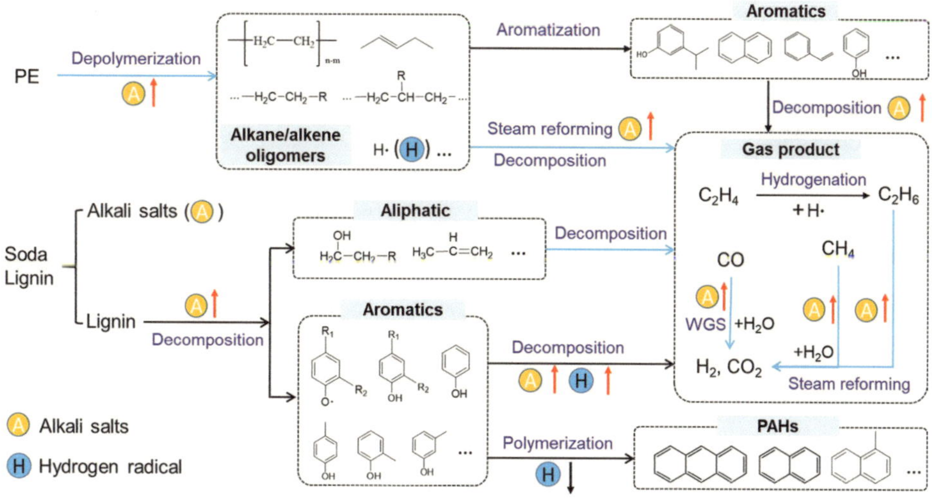

Figure 7.
Interaction mechanism of alkali lignin and PE in supercritical water gasification [64].

An integrated system of SCWG of black liquor coupling with pulping process to mainly produce hydrogen as well as electric power and steam for the pulping plant was proposed in our previous study (**Figure 8**) [55]. The high-temperature and high-pressure gas products were used to produce power and steam through steam turbines and heat exchangers. Hydrogen was purified through PSA, and the residual gas and supplementary natural gas were burned to provide heat for SCWG reaction. In this study, a pulp mill with a daily output of 1000 ADt (Air Dry ton) pulp was selected as a reference, and the main data about the energy consumption were collected from the literatures [65, 66]. The reaction temperature and pressure of SCWG were set at 700°C and 25 MPa, respectively. From this integration system, 37,126 Nm^3/h high-purity hydrogen, 53,053 kg/h MP steam, and 72,446 kg/h LP steam can be produced. The generated MP and LP steam could fully meet the requirements of the pulping process, and some excess heating could also be exported to the nearby community. However, some electric power needs to be imported to fulfill the requirements of the plant. The difference between using air and oxygen as oxidants in the burner of natural gas and exhaust gas products was evaluated and found that using air was more energy-efficient, but the introduction of nitrogen would generate pollutants (NO_x), and the tail gas needs to be treated. The comprehensive energy consumption of pulp production from the pulping process coupling with SCWG of black liquor was calculated. For 1 ADt pulp produced, 221.48 kgce energy can be produced by using air as oxidant, and 288.01 kgce energy was consumed when using oxygen as oxidant. These values were much lower than the energy consumption level of China for pulp production in 2015 (approximately 370 kgce/ADt) [67].

An advantage of SCWG over conventional treatment is that black liquor can be gasified directly without energy-intensive evaporation. However, concentrating black liquor to a certain extent can reduce the system scale and related heating loss. To balance the evaporation energy consumption and heating loss reduction, the necessity of black liquor evaporation was evaluated based on energy and exergy analysis of the above system (**Figure 8**). This evaluation used the steam and power generated from SCWG of black liquor for evaporation to improve the energy efficiency. In different pulping processes, the concentration of weak black liquor varied in the range of

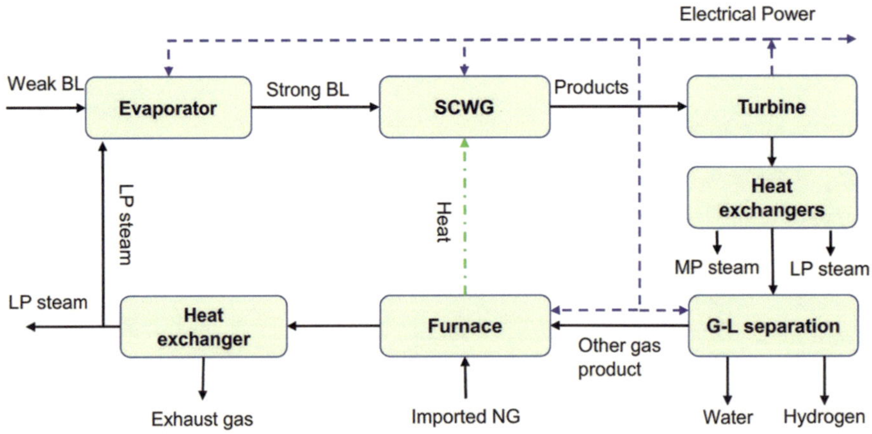

Figure 8.
Conceptual diagram of SCWG of black liquor coupled with evaporation. The green dot dash line and blue dashed lines represent the heat and power flow between the units, respectively [55].

10–20 wt% for different pulping processes [10]. And the influence of this concentration on the system efficiency was studied firstly [56]. With weak black liquor concentration increasing from 10 wt% to 20 wt%, the energy efficiency decreased, and the exergy efficiency increased. When coupled with black liquor evaporation and the concentration was increased from 15 wt% to 40 wt%, the energy efficiency decreased due to energy consumption of evaporation, and the exergy efficiency first increased and then decreased. The maximum exergy efficiency of 41.95% was obtained when the black liquor concentration was 21.82 wt% (optimal concentration). For the black liquor with different initial concentrations, the influence of evaporation was also different. The dilute black liquor needed to evaporate more water to reach the optimal concentration, and the efficiency improvement caused by evaporation was greater. The effect number of the multieffect evaporators was also studied, and it was found that the optimal concentration increased from 19.93 wt% to 23.13 wt% when the effect number increased from 4 to 7.

Some scientists noticed the high energy loss in hydrogen separation by PSA and used the syngas chemical looping (SCL) coupled with SCWG of black liquor to improve the energy efficiency. Taking advantage of the SCL to produce H_2 and CO_2, it can not only produce hydrogen and power but also avoid additional CO_2 separation processes. Ajiwibowo et al. [68] designed a system composed of SCWG of black liquor, SCL, and power generation process to produce hydrogen and power (**Figure 9**). SCL was used to reduce the syngas product by Fe_2O_3. The generated steam and CO_2 were used to produce power by steam turbine, and the generated Fe and FeO were oxidized by steam to generate hydrogen and Fe_3O_4. The steam and hydrogen flows were also used to produce power by steam turbine, and high-purity hydrogen could be separated after cooling. Then Fe_3O_4 reacted with oxygen in a furnace to regenerate Fe_2O_3, which released heat for SCWG reaction. The effects of temperature (375, 500, and 600°C) on gasification products and system efficiency were studied. The highest energy efficiency of 73% was obtained at 600°C. At higher temperature, the concentration of hydrogen was lower, while the concentrations of CO, C_2H_6, and CH_4 were higher. Darmawan et al. [69] further discussed this system. The maximum efficiency was 82%, which was higher than that of traditional gasification system (69.1%), and the CO_2 capture of the system was 75%. In addition, the influence of pressure in

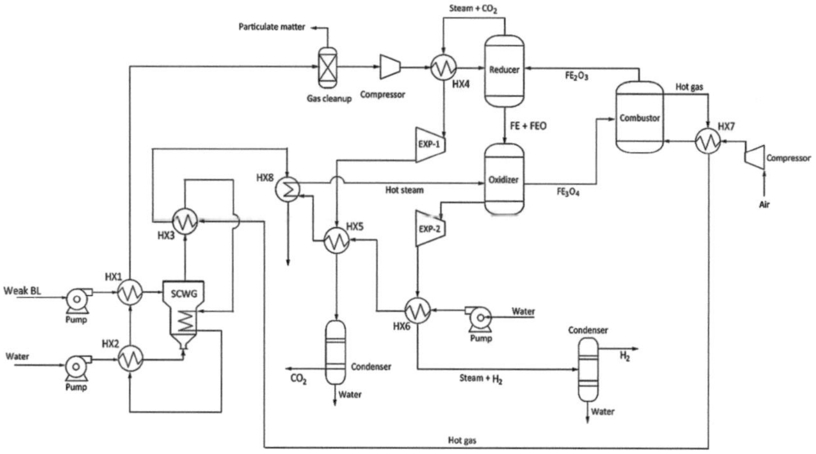

Figure 9.
Schematic diagram of SCWG of BL, chemical looping, and power generation [68].

SCL was also studied. When the pressure increased from 2.5 MPa to 3.5 MPa, there is no significant change in energy efficiency, but the consumed power in pump and compressor increased and the net power generation was decreased.

To evaluate the impact of SCWG of black liquor on pulping process in terms of raw material recovery, energy utilization, and economic benefits, Magdeldin et al. [70] designed an integrated process of SCWG and Nordic pulp mills. In this study, the process of combining SCWG with recovery boiler was proposed, and the influence of SCWG product on the performance of the recovery boiler was studied. The energy efficiency of SCWG system reached 83% and 80% at 450°C and 600°C, respectively, which was basically consistent with above studies [55, 56]. The gas products were used as the alternative fuel of heavy fuel oil in causticizing plant, and the surplus was sold to the public society. They also studied the impact of integrating SCWG system on the capacity of pulping plant and found that the pulping capacity can be increased by 75% on the premise of meeting the power and steam demand of the pulping plant. If it is considered to purchase power from outside and separate 75–77% of the black liquor into the SCWG system, the pulping capacity can be tripled. The economic benefits of the integrated gasification system were calculated. Compared with the reference Nordic pulp mill, the lowest selling price per 1ADt pulp can be reduced by 22%.

To explore the technoeconomic feasibility of SCWG of black liquor in sulfate pulp mills, Özdenkçi et al. [15] evaluated different integrated schemes using different reactor materials for hydrogen, heat, and power generation. Compared with stainless steel, the use of Inconel reactor was found to be able to obtain better economic benefits and reduce the cost of hydrogen production. Subsequently, Özdenkçi et al. [71] studied the effects of different process conditions. The main process conditions included: stainless steel and Inconel reactor, temperature (600–750°C), short (133–162 s) and long (300 s) residence time. It showed that using Inconel reactor helped achieve the highest cold gas efficiency (87.8%) and hydrogen yield (24.92 mol/kg) at 750°C with a long residence time (300 s). Through the technical and economic evaluation, it was concluded that the longer residence time at 700°C and the shorter residence time at 750°C could be used as alternative conditions.

7. Conclusions

SCWG is an innovative and potential treatment method for black liquor from pulping processes. Several researchers have studied this topic, and many progresses had been made, such as:

1. SCWG is a promising handling method for black liquor, where both hydrogen production and pollution handling can be realized simultaneously. After treatment at 600°C, the COD concentration decreased from 95,000 mg/L to 2160 mg/L. The increase in temperature enhanced gasification, and the highest carbon gasification efficiency of 94.10% and hydrogen fraction of nearly 70% were achieved at 750°C. The inherent alkali, prolongation of reaction time, diluting the black liquor, and the use of Inconel-625 reactor favored gasification.

2. Metal oxide was a better choice to be used as catalyst for SCWG of black liquor than alkali and metal catalysts due to the alkali and sulfur content of black liquor. With the catalysis of metal oxide, the oxygen species can promote black liquor

degradation, and the reduced metal can also serve as the catalyst. The combination of different metal oxides, such as CuO-ZnO, CeO_2-ZrO_2, and Fe_2O_3-Cr_2O_3, can further improve catalyst activity.

3. Some accessible energy sources, including coal, biomass, and plastics, can be co-gasified with black liquor to produce more energy for pulp mill to realize the energy self-sufficiency. The synergistic effect was found in their co-gasification because the alkali in black liquor can promote the gasification of other substances, and the addition of other substances can take more fully use of the abundant alkali in black liquor.

4. The utilization of hot-compressed products of SCWG of black liquor is the main optimization direction of the integrated system. At present, the main utilization methods in the system analysis include: PSA separation of hydrogen, production of hydrogen and power by chemical looping, as alternative fuel in causticizing plants. Considering the difficulties in the storage and transportation of hydrogen and the energy consumption in the separation process, CHP is a more practical utilization method that should be further analyzed and optimized.

Though great progress has been made in SCWG of black liquor, several challenges still need to be handled before its industrial utilization. For example, reactor corrosion can be affected by the operating parameters, reactor materials, and properties of the feedstock. The alkali salts in black liquor were found to be able to aggravate the reactor corrosion. In addition, reactor plugging was another challenge to be overcome for continuous running of the system. Both the char generated for incomplete gasification and alkali salt precipitation in SCW can result in reactor plugging. Furthermore, though the distribution of the alkali salts in SCWG was revealed, the development of the proper method and equipment for the alkali recovery was still needed to improve the economic efficiency of this technology.

Acknowledgements

The authors are grateful for the financial support from Fundamental Research Funds for the Central Universities (xzy012020072).

Author details

Changqing Cao[1,2]*, Lihui Yu[2], Wenhao Li[2], Lanjun Liu[2] and Peigao Duan[1]*

1 School of Chemical Engineering and Technology, Xi'an Jiaotong University, Xi'an, China

2 State Key Laboratory of Multiphase Flow in Power Engineering, Xi'an Jiaotong University, Xi'an, China

*Address all correspondence to: cq.cao@mail.xjtu.edu.cn and pgduan@xjtu.edu.cn

References

[1] Sricharoenchaikul V. Assessment of black liquor gasification in supercritical water. Bioresource Technology. 2009;**100**(2):638-643

[2] Verma OP, Manik G, Sethi SK. A comprehensive review of renewable energy source on energy optimization of black liquor in MSE using steady and dynamic state modeling, simulation and control. Renewable and Sustainable Energy Reviews. 2019;**100**:90-109

[3] Vähä-Savo N, DeMartini N, Hupa M. Combustion of black liquor–solid biomass mixtures in a single particle reactor characteristics and fate of nitrogen. Energy & Fuels. 2011;**25**(11):4944-4951

[4] Demirbaş A. Pyrolysis and steam gasification processes of black liquor. Energy Conversion and Management. 2002;**43**(7):877-884

[5] Caravaca A, Garcia-Lorefice WE, Gil S, de Lucas-Consuegra A, Vernoux P. Towards a sustainable technology for H_2 production: Direct lignin electrolysis in a continuous-flow polymer electrolyte membrane reactor. Electrochemistry Communications. 2019;**100**:43-47

[6] Rodriguez Correa C, Kruse A. Supercritical water gasification of biomass for hydrogen production—Review. The Journal of Supercritical Fluids. 2017;**S0896844617303510**

[7] Cao C, Xu L, He Y, Guo L, Jin H, Huo Z. High-efficiency gasification of wheat straw black liquor in supercritical water at high temperatures for hydrogen production. Energy & Fuels. 2017;**31**(4):3970-3978

[8] Boucard H, Watanabe M, Takami S, Weiss-Hortala E, Barna R, Adschiri T. Beneficial use of CeO_2 nanocatalyst for black liquor conversion under sub and supercritical conditions. The Journal of Supercritical Fluids. 2015;**105**:66-76

[9] Cao C, Guo L, Chen Y, Guo S, Lu Y. Hydrogen production from supercritical water gasification of alkaline wheat straw pulping black liquor in continuous flow system. International Journal of Hydrogen Energy. 2011;**36**(21):13528-13535

[10] Joelsson J, Gustavsson L. CO2 emission and oil use reduction through black liquor gasification and energy efficiency in pulp and paper industry. Resources, Conservation and Recycling. 2008;**52**(5):747-763

[11] Cao C, Xie Y, Mao L, Wei W, Shi J, Jin H. Hydrogen production from supercritical water gasification of soda black liquor with various metal oxides. Renewable Energy. 2020;**157**:24-32

[12] Casademont P, Sánchez-Oneto J, Scandelai A, Cardozo-Filho L, Portela J. Hydrogen production by supercritical water gasification of black liquor: Use of high temperatures and short residence times in a continuous reactor. The Journal of Supercritical Fluids. 2020;**159**:104772

[13] Ryczkowski R, Goscianska J, Panek R, Franus W, Przybysz K, Grams J. Sustainable nickel catalyst for the conversion of lignocellulosic biomass to H2-rich gas. International Journal of Hydrogen Energy. 2021;**46**(18):10708-10722

[14] De Blasio C, Lucca G, Özdenkci K, Mulas M, Lundqvist K, Koskinen J, et al. A study on supercritical water gasification of black liquor conducted in stainless steel and

nickel-chromium-molybdenum reactors. Journal of Chemical Technology & Biotechnology. 2016;**91**(10):2664-2678

[15] Özdenkçi K, De Blasio C, Sarwar G, Melin K, Koskinen J, Alopaeus V. Techno-economic feasibility of supercritical water gasification of black liquor. Energy. 2019;**189**:116284

[16] Cao C, Guo L, Yin J, Jin H, Cao W, Jia Y, et al. Supercritical water gasification of coal with waste black liquor as inexpensive additives. Energy & Fuels. 2015;**29**(1):384-391

[17] Hawangchu Y, Atong D, Sricharoenchaikul V. The effect of alkali on the product distribution from black liquor conversion under supercritical water. Environmental Technology. 2017;**38**(13-14):1742-1750

[18] Rönnlund I, Myréen L, Lundqvist K, Ahlbeck J, Westerlund T. Waste to energy by industrially integrated supercritical water gasification–Effects of alkali salts in residual by-products from the pulp and paper industry. Energy. 2011;**36**(4):2151-2163

[19] Järvinen MP, Kankkunen A, De Blasio C, Miikkulainen P. Vapor pressure and boiling point elevation of black liquor. Nordic Pulp & Paper Research Journal. 2015;**30**(3):411-416

[20] Andersson E, Harvey S. System analysis of hydrogen production from gasified black liquor. Energy. 2006;**31**(15):3426-3434

[21] Lappalainen J, Baudouin D, Hornung U, Schuler J, Melin K, Bjelić S, et al. Sub-and supercritical water liquefaction of kraft lignin and black liquor derived lignin. Energies. 2020;**13**(13):3309

[22] Cao C, Guo L, Jin H, Guo S, Lu Y, Zhang X. The influence of alkali precipitation on supercritical water gasification of glucose and the alkali recovery in fluidized-bed reactor. International Journal of Hydrogen Energy. 2013;**38**(30):13293-13299

[23] Fedyaeva ON, Vostrikov AA, Shishkin AV, Dubov DY. Conjugated processes of black liquor mineral and organic components conversion in supercritical water. The Journal of Supercritical Fluids. 2019;**143**:191-197

[24] Elliott DC. Catalytic hydrothermal gasification of biomass. Biofuels, Bioproducts and Biorefining. 2008;**2**(3):254-265

[25] Okolie JA, Rana R, Nanda S, Dalai AK, Kozinski JA. Supercritical water gasification of biomass: A state-of-the-art review of process parameters, reaction mechanisms and catalysis. Sustainable Energy & Fuels. 2019;**3**(3):578-598

[26] Sinağ A, Kruse A, Rathert J. Influence of the heating rate and the type of catalyst on the formation of key intermediates and on the generation of gases during hydropyrolysis of glucose in supercritical water in a batch reactor. Industrial and Engineering Chemistry Research. 2004;**43**(2):502-508

[27] Onwudili JA, Williams PT. Role of sodium hydroxide in the production of hydrogen gas from the hydrothermal gasification of biomass. International Journal of Hydrogen Energy. 2009;**34**(14):5645-5656

[28] Sheikhdavoodi MJ, Almassi M, Ebrahimi-Nik M, Kruse A, Bahrami H. Gasification of sugarcane bagasse in supercritical water; evaluation of alkali catalysts for maximum hydrogen production. Journal of the Energy Institute. 2015;**88**(4):450-458

[29] Guo S, Guo L, Cao C, Yin J, Lu Y, Zhang X. Hydrogen production from glycerol by supercritical water gasification in a continuous flow tubular reactor. International Journal of Hydrogen Energy. 2012;**37**(7):5559-5568

[30] Hupa M. Recovery boiler chemical principles. TAPPI Kraft Recovery Course 2007;**2007**:2

[31] Savage PE. A perspective on catalysis in sub-and supercritical water. The Journal of Supercritical Fluids. 2009;**47**(3):407-414

[32] Okolie JA, Mukherjee A, Nanda S, Dalai AK, Kozinski JA. Catalytic supercritical water gasification of soybean straw: Effects of catalyst supports and promoters. Industrial & Engineering Chemistry Research. 2021;**60**(16):5770-5782

[33] Peng G, Ludwig C, Vogel F. Catalytic supercritical water gasification: Interaction of sulfur with ZnO and the ruthenium catalyst. Applied Catalysis B: Environmental. 2017;**202**:262-268

[34] Osada M, Hiyoshi N, Sato O, Arai K, Shirai M. Effect of sulfur on catalytic gasification of lignin in supercritical water. Energy & Fuels. 2007;**21**(3):1400-1405

[35] Osada M, Hiyoshi N, Sato O, Arai K, Shirai M. Reaction pathway for catalytic gasification of lignin in presence of sulfur in supercritical water. Energy & Fuels. 2007;**21**(4):1854-1858

[36] Zhan X, Zhou Z, Wang F. Catalytic effect of black liquor on the gasification reactivity of petroleum coke. Applied Energy. 2010;**87**(5):1710-1715

[37] Azadi P, Afif E, Azadi F, Farnood R. Screening of nickel catalysts for selective hydrogen production using supercritical water gasification of glucose. Green Chemistry. 2012;**14**(6):1766-1777

[38] Al'Myasheva O, Korytkova E, Maslov A, Gusarov V. Preparation of nanocrystalline alumina under hydrothermal conditions. Inorganic Materials. 2005;**41**(5):460-467

[39] Zhang Y, Li Y, Gu J, Tian S, Ning P. Hydrothermal stability of different zeolites in supercritical water: Implication for synthesis of supported catalysts by supercritical water impregnation. Korean Journal of Chemical Engineering. 2018;**35**(9):1932-1940

[40] Liu D, Chen S, Fei X, Huang C, Zhang Y. Regenerable CuO-based adsorbents for low temperature desulfurization application. Industrial & Engineering Chemistry Research. 2015;**54**(14):3556-3562

[41] Ates A, Azimi G, Choi K-H, Green WH, Timko MT. The role of catalyst in supercritical water desulfurization. Applied Catalysis B: Environmental. 2014;**147**:144-155

[42] Oh W-D, Lei J, Veksha A, Giannis A, Chan W-P, Lisak G, et al. Ni-Zn-based nanocomposite loaded on cordierite mullite ceramic for syngas desulfurization: Performance evaluation and regeneration studies. Chemical Engineering Journal. 2018;**351**:230-239

[43] Park KC, Tomiyasu HJCC. Gasification reaction of organic compounds catalyzed by RuO_2 in supercritical water. Chemical Communications. 2003;**6**:694-695

[44] Watanabe M, Inomata H, Arai K. Catalytic hydrogen generation from biomass (glucose and cellulose) with ZrO_2 in supercritical water. Biomass and Bioenergy. 2002;**22**(5):405-410

[45] Seif S, Fatemi S, Tavakoli O, Bahmanyar H. Hydrogen production through hydrothermal gasification of industrial wastewaters using transition metal oxide catalysts. The Journal of Supercritical Fluids. 2016;**114**:32-45

[46] Cao C, Zhang Y, Cao W, Jin H, Guo L, Huo Z. Transition metal oxides as catalysts for hydrogen production from supercritical water gasification of glucose. Catalysis Letters. 2017;**147**(4):828-836

[47] Gong W, Zhou Z, Liu Y, Wang Q, Guo L. Catalytic gasification of sewage sludge in supercritical water: Influence of K_2CO_3 and H_2O_2 on hydrogen production and phosphorus yield. ACS Omega. 2020;**5**(7):3389-3396

[48] Hao X, Guo L, Zhang X, Guan Y. Hydrogen production from catalytic gasification of cellulose in supercritical water. Chemical Engineering Journal. 2005;**110**(1-3):57-65

[49] Yu J, Lu X, Shi Y, Chen Q, Guan Q, Ning P, et al. Catalytic gasification of lignite in supercritical water with Ru/CeO2–ZrO2. International Journal of Hydrogen Energy. 2016;**41**(8):4579-4591

[50] Kirszensztejn P, Przekop R, Tolińska A, Maćkowska E. Pyrolytic and catalytic conversion of rape oil into aromatic and aliphatic fractions in a fixed bed reactor on Al_2O_3 and Al_2O_3/B_2O_3 catalysts. Chemical Papers. 2009;**63**(2):226

[51] Duman G, Watanabe T, Uddin MA, Yanik J. Steam gasification of safflower seed cake and catalytic tar decomposition over ceria modified iron oxide catalysts. Fuel Processing Technology 2014;**126**:276-283

[52] Mastuli MS, Kamarulzaman N, Kasim MF, Sivasangar S, Saiman MI, Taufiq-Yap YH. Catalytic gasification of oil palm frond biomass in supercritical water using MgO supported Ni, Cu and Zn oxides as catalysts for hydrogen production. International Journal of Hydrogen Energy. 2017;**42**(16):11215-11228

[53] Cao C, Xie Y, Li L, Wei W, Jin H, Wang S, et al. Supercritical water gasification of lignin and cellulose catalyzed with co-precipitated CeO2–ZrO2. Energy & Fuels. 2021;**35**(7):6030-6039

[54] Cao C, Xie Y, Chen Y, Lu J, Shi J, Jin H, et al. Hydrogen production from supercritical water gasification of lignin and cellulose with coprecipitated CuO–ZnO and Fe2O3–Cr2O3. Industrial & Engineering Chemistry Research. 2021;**60**(19):7033-7042

[55] Cao C, Guo L, Jin H, Cao W, Jia Y, Yao X. System analysis of pulping process coupled with supercritical water gasification of black liquor for combined hydrogen, heat and power production. Energy. 2017;**132**:238-247

[56] Cao C, He Y, Chen J, Cao W, Jin H. Evaluation of effect of evaporation on supercritical water gasification of black liquor by energy and exergy analysis. International Journal of Hydrogen Energy. 2018;**43**(30):13788-13797

[57] Wang L. Sea lanes and Chinese national energy security. Journal of Coastal Research. 2015;**73**(10073):572-576

[58] Guo L, Jin H, Ge Z, Lu Y, Cao C. Industrialization prospects for hydrogen production by coal gasification in supercritical water and novel thermodynamic cycle power generation system with no pollution emission. Science China Technological Science. 2015;**58**(12):1989-2002

[59] Ge Z, Guo S, Guo L, Cao C, Su X, Jin H. Hydrogen production by non-catalytic partial oxidation of coal in supercritical water: Explore the way to complete gasification of lignite and bituminous coal. International Journal of Hydrogen Energy. 2013;**38**(29):12786-12794

[60] Kershaw JR. Comments on the role of the solvent in supercritical fluid extraction of coal. Fuel. 1997;**76**:453-454

[61] Cao C, He Y, Wang G, Jin H, Huo Z. Co-gasification of alkaline black liquor and coal in supercritical water at high temperatures (600-750°C). Energy & Fuels. 2017;**31**(12):13585-13592

[62] Cao C, Zhang Y, Li L, Wei W, Wang G, Bian C. Supercritical water gasification of black liquor with wheat straw as the supplementary energy resource. International Journal of Hydrogen Energy. 2019;**44**(30):15737-15745

[63] Tejaswini M, Pathak P, Ramkrishna S, Ganesh SP. A comprehensive review on integrative approach for sustainable management of plastic waste and its associated externalities. Science of the Total Environment. 2022;**825**:153973

[64] Cao C, Bian C, Wang G, Bai B, Xie Y, Jin H. Co-gasification of plastic wastes and soda lignin in supercritical water. Chemical Engineering Journal. 2020;**388**:124277

[65] Naqvi M, Yan J, Dahlquist E. Synthetic gas production from dry black liquor gasification process using direct causticization with CO2 capture. Applied Energy. 2012;**97**:49-55

[66] Andersson E, Harvey S. System analysis of hydrogen production from gasified black liquor. Energy. 2006;**31**(15):3426-3434

[67] Xue J, Zhao Z, Dai Y, Wang B. Green Low-Carbon Development in China. Springer; 2013

[68] Ajiwibowo MW, Darmawan A, Aziz M. Co-production of hydrogen and power from black liquor via supercritical water gasification, chemical looping and power generation. Energy Procedia. 2019;**158**:2299-2304

[69] Darmawan A, Ajiwibowo MW, Biddinika MK, Tokimatsu K, Aziz M. Black liquor-based hydrogen and power co-production: Combination of supercritical water gasification and syngas chemical looping. Applied Energy. 2019;**252**:113446

[70] Magdeldin M, Järvinen M. Supercritical water gasification of Kraft black liquor: Process design, analysis, pulp mill integration and economic evaluation. Applied Energy. 2020;**262**:114558

[71] Özdenkçi K, Prestipino M, Björklund-Sänkiaho M, Galvagno A, De Blasio C. Alternative energy valorization routes of black liquor by stepwise supercritical water gasification: Effect of process parameters on hydrogen yield and energy efficiency. Renewable and Sustainable Energy Reviews. 2020;**134**:110146

Chapter 8

Minimising CO_2 Emissions from Coal Gasification

Shaakirah Cassim and Shehzaad Kauchali

Abstract

Traditional coal-to-liquid processes use gasification with excess steam to obtain hydrogen-rich syngas for downstream manufacturing of methanol or Fischer-Tropsch liquids. Such processes are shown to produce very large amounts of CO_2 directly by the Water-Gas-Shift (WGS) reaction or, indirectly, by combustion in raising steam. It is shown how any coal gasifier can operate under auto-thermal conditions with methane as source of hydrogen instead of steam. This co-gasification system produces syngas for a poly-generation facility while minimising the formation of process CO_2. It is shown that minimal steam is required for the process and a limit on the maximum amount of H_2:CO can be obtained. Co-gasification of coal is shown to have a major advantage in that a separate WGS reactor is not required, less CO_2 is formed and methane is reformed non-catalytically within the gasification unit. Furthermore, regions of thermally balanced operations were identified that enabled a targeting approach for the design of co-gasification systems. The method will guide gasification practitioners to incorporate fossil fuels and renewable-H_2 into coal-to-liquids processes that require syngas with H_2:CO ratio of 2. An important result shows that low-grade coals can be co-gasified with methane to obtain CO_2-free syngas ideal for power generation.

Keywords: *CHO*-diagram, coal-to-liquid, CO_2, Co-gasification, renewable hydrogen

1. Introduction

A major concern with coal-to-liquids (CTL) producing facilities is the unprecedented amount of carbon dioxide they produce. Efforts to capture and sequester this unwanted green-house gas in geological formations and, or, for enhanced oil recovery activities are encouraging albeit at a penalty cost to overall plant efficiencies and economics. For energy security in geographically stranded economies, coal and methane (either from unconventional sources such as coal-bed methane or shale gas) will play an integral role in future energy developments and what remains is the planning and execution of such activities with an environmentally conscience philosophy until carbon-free economies are the mainstay. Williams [1] suggested that a key enabling strategy leading to attractive energy-costs, without further technological developments, is "polygeneration" defined as co-production from synthesis gas of at least electricity and one or more clean synthetic fuels such as Fischer-Tropsch (FT) liquids, methanol and dimethyl ether (DME). The advantage of polygeneration is to aid a wide range of energy needs with extremely low levels of emissions, often higher efficiencies and lower cost [1].

A key step of polygeneration facilities, using coal, is the production of syngas, comprising mainly of hydrogen, carbon monoxide, carbon dioxide, methane and steam. For these types of facilities to coexist, where the target is "clean" gas enriched with hydrogen and carbon monoxide only, a gasifier that operates as "partial combustion" of coal is required. For power generation this syngas is then fed towards an integrated combined cycle electricity power block comprising of a gas turbine and steam turbine system as in an integrated gasification combined cycle (IGCC) process. However, for liquid-fuels production such as for methanol or FT as shown by Battaerd & Evans [2], the syngas is corrected for its H_2:CO ratio using an additional equilibrium-limited water-gas-shift (WGS) reactor and with excess steam to drive the reaction towards an increase in H_2-content. There are thus two main, undesirable, effects of the addition of the WGS reactor: firstly, the WGS reaction itself creates CO_2 and, secondly, large amounts of steam is needed for favourable equilibrium necessitating some of the original coal (or tail gases) to be combusted leading to further creation of process CO_2 emitted to the atmosphere. This phenomenon is also noted in some CTL processes where traditional fixed-bed counter current operations achieve simultaneous gasification of coal (with excess steam) and the correction of the H_2:CO ratio in a single piece of equipment [2]. An important strategy, in limiting the amount of CO_2 produced in coal-gasification processes, is thus to avoid the phenomenon of the WGS reaction by restricting the amount of steam used.

Another strategy to correct the H_2:CO ratio as used by FT processes, described in the works of Probstein & Hicks [3], is the mixing of hydrogen-rich syngas recycled from the autothermal steam reforming of tail gases comprising primarily of methane. However, this is generally acceptable practice if the initial syngas product from the gasification process has a high methane content and the FT catalyst itself produces a significant amount of methane by-product. It is an opinion that the highly inefficient autothermal steam reforming process requires a large amount of excess steam for equilibrium and also produces a large amount of CO_2 due to the WGS reaction (occurring simultaneously with gasification) and indirectly from combustion in raising the steam.

Steam reforming of methane is the predominant method for the production of syngas at industrial scale. Cao et al. [4] note that natural gas based syngas are capital intensive due to expensive catalysts used and often are associated with higher energy consumption. There is thus a drive for the development of alternative technology for cost-effective production of syngas gas using geographically abundant and cheap feedstock such as coal. It is noted that there are challenges in decreasing capital investment and operational cost of coal based syngas process with flexible H_2/CO ratios [4]. Firstly, coal gasification leads to low H_2:CO ratios and secondly the process economics is strongly affected by the coal reactivity as this determines the carbon conversion and gas yields. Furthermore Wu & Wang [5] identified that methane could be an ideal source for H_2, for syngas conversion requiring high H_2, and that coal-bed gas is a good methane source for co-gasification purposes. Lastly, a co-gasification experiment in a fluidized bed was performed to study the effects of adjusting the methane amount on the H_2/CO ratio. Wu & Wang [5] performed similar experiments with bituminous coal and anthracite to demonstrate the combined coal gasification and methane reforming process in a single reactor. One of the objectives in the works of Wu & Wang [5] was to elucidate the catalytic effect of the unreacted coal char and ash on the partial oxidation or steam reforming of natural gas in the fluidized bed operating at 1000°C. Syngas comprising of H_2:CO ratio of 1 was achieved with significant amounts of CO_2 in the product. Song & Guo [6] suggest a co-gasification experiment in a moving-bed configuration using a modified large-volume blast furnace with lime-containing liquid-flux for absorption of sulphur compounds. They noted that the theoretical H_2/CO ratio could vary between 0.4 (coal gasification) and 2 (partial oxidation of methane) within their system. It was experimentally observed that the H_2:CO ratio was dependent on the O_2/CH_4 ratio in the feed. For O_2/CH_4 ratio in the feed below 1, the H_2/CO in the product syngas is greater than 1 with over 90% gas being H_2 and CO. Ouyang et al. [7] validate the need to achieve endothermic and exothermic reactions in a single reactor stating the advantages of the co-gasification of coal with methane as follows: low production cost of syngas, adjustable H_2/CO ratio in range 1–2, high steam and methane conversion, energy savings and flexibility in using various carbon containing feedstock.

The work by Kauchali [8] presents an interesting theoretical basis for the analysis of coal gasification process using bond equivalent diagrams developed by Battaerd & Evans [2] for elemental carbon. Here, it was shown that theoretical gasification thermally-balanced regions could be obtained purely by analysis of the basic stoichiometry of the coal-oxidants system. The results in [8] showed that real coal gasification systems operated in, or close to the regions, predicted theoretically, and that the method proved to be an indispensable tool in understanding underground coal gasification processes.

The co-gasification of coal with methane developed above, in principle, to produce syngas with a high H_2:CO ratios rely on the fact that the partial combustion of coal is highly exothermic driving the endothermic steam reforming of methane in an autothermal and balanced manner. Unfortunately, this is only true for high grade coal with high calorific values (CV in MJ/kg). In this paper it will be shown, for a typical South African coals with low CV (bituminous and sub-bituminous), that certain critical co-feed conditions (amounts of CH_4 and coal) are required to be met, to achieve a CO_2-free syngas, that can be used in a polygeneration facility irrespective of the gasifier type and flow configurations. In addition, the minimum H_2:CO ratio achievable for the thermally balanced co-gasification of SA coal will be determined – this limit will ultimately determine the minimum amount of renewable hydrogen needed for supplementation of the syngas for liquids production. Fundamentally, the need for steam in the gasification process, as a source of hydrogen, is obviated and the endothermic partial oxidation of coal is a practical way for temperature control on the limit of flame temperature in the combustion zone.

2. Analysis of coal-methane Co-gasification

A systematic method of obtaining the stoichiometric reactions and thermally balanced region for the gasification of a South African coal from Bosjesspruit mine is studied. A carbon-hydrogen-oxygen (CHO) ternary or bond equivalent diagram is used for this type of analysis. A material balance, thermodynamic equilibrium as well as insight on the product composition of the gasification process can be accomplished with the ternary diagram. The feed and product highways are used to determine the stoichiometric reactions, indicated by the intersections, for the gasification process.

Thereafter, the thermal nature of the reactions (endothermic or exothermic) can be used to determine the thermally balanced region of operation as done in other previous studies.

2.1 Ternary bond equivalent diagram

The composition of coal, indicated by coal composition charts, can be represented by three atomic species, namely, carbon (C), hydrogen (H) and oxygen (O). Hence, a ternary (CHO) graphical illustration can be used to represent the gasification reactions [2]. The ternary diagram is also referred to as a bond equivalent (BE) phase diagram. The diagram comprises of an equilateral triangular grid with lines parallel to the three sides drawn within the triangle and C, H and O on each vertex (see **Figure 1**). The vertex thus represents the pure component i.e., 100% C, H or O. The edges of the triangle represent a binary mixture of the atomic species. For example, a point alone the C-H edge will represent a species composed of C and H only (without any O). Furthermore, any point within the triangle will represent a mixture of the three atomic species with different compositions [9].

The bonding capacity of the constituent elements is used to denote the species. For a species $C_xH_yO_z$, the bond equivalent compositions can be calculated using bonding capacity of each element and normalised accordingly.

2.2 Feed and product highways

The feed and product components and highways are plotted in **Figure 2**. Methane (CH_4) is represented by the point between 100% C and 100% H such that C and H have combined to achieve their normal valencies. The point midway between 100% H and 100% O represents water (H_2O) and the point midway between 100% C and

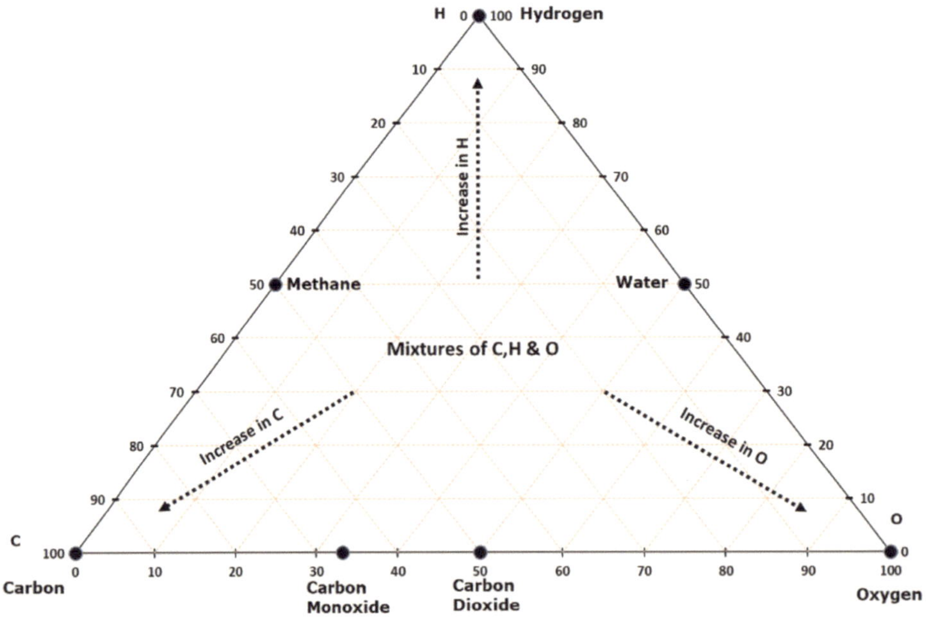

Figure 1.
Representation of a CHO ternary diagram [9].

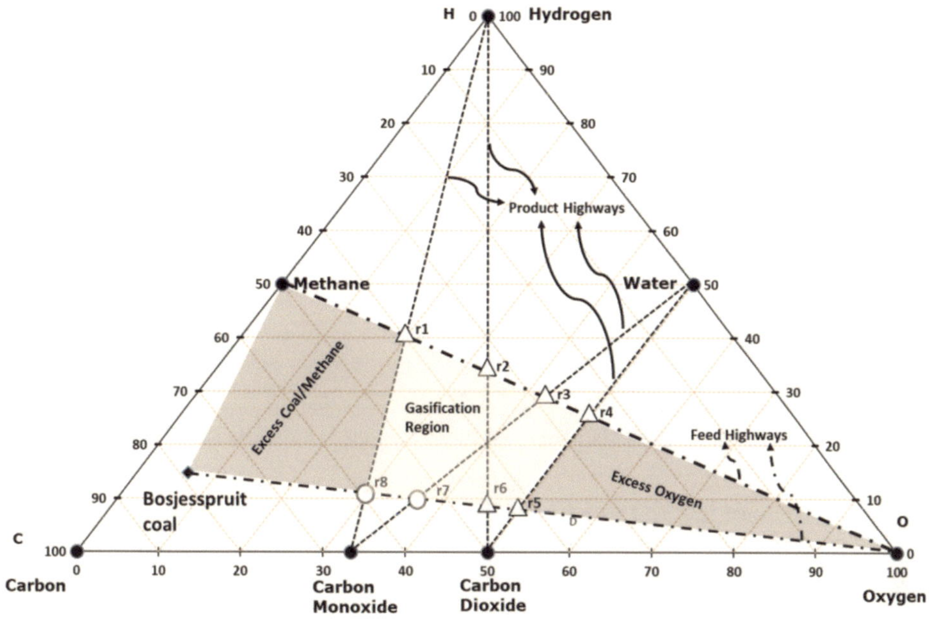

Figure 2.
Feed and product highways for co-gasification of Bosjesspruit coal with methane.

100% O represents carbon dioxide (CO_2) [2]. In this manner, using the BE composition calculations, the components are plotted. The Ultimate Analysis of the Bosjesspruit coal determines its molecular formula to be $CH_{0.75}O_{0.16}$ [8]. From the Proximate Analysis % [8]: moisture is 3.9, Ash is 32.8, Volatile Matter is 21.6 and Fixed Carbon is 52.2. The calorific value of the coal, as received, is 18.88 MJ/kg rendering it a low quality sub-bituminous coal.

The following calculations are used for the BE composition of the coal:

$$C = \frac{4(4)}{4(4) + 1(0.75) + 2(0.16)} = 0.79 \tag{1}$$

$$H = \frac{1(0.75)}{4(4) + 1(0.75) + 2(0.16)} = 0.15 \tag{2}$$

$$O = \frac{2(0.16)}{4(4) + 1(0.75) + 2(0.16)} = 0.06 \tag{3}$$

Additionally, the feed and product components are joined by the lines referred to as the feed and product highways. For example, the feed highways are the lines connecting coal to oxygen and methane to oxygen. The product highways are connected from H_2 to CO/CO_2 and H_2O to CO/CO_2 points.

2.3 Representation of important reactions for Co-gasification

The intersections between these highways represent the important gasification reaction points (stoichiometric reactions) as summarised in **Table 1**. The triangular points, in **Figure 2**, indicate the exothermic reactions (r1 to r6) and the circled points are the two endothermic reactions (r7 & r8).

No.	Reaction	Heat of reaction (kJ/mol)
r1	$CH_4 + 0.5O_2 \rightarrow CO + 2H_2$	−35.50 (exothermic)
r2	$CH_4 + O_2 \rightarrow CO_2 + 2H_2$	−318.70 (exothermic)
r3	$CH_4 + 1.5O_2 \rightarrow CO + 2H_2O$	−519.10 (exothermic)
r4	$CH_4 + 2O_2 \rightarrow CO_2 + 2H_2O$	−802.30 (exothermic)
r5	$CH_{0.75}O_{0.16} + 1.12O_2 \rightarrow CO_2 + 0.375H_2O$	−272.53 (exothermic)
r6	$CH_{0.75}O_{0.16} + 0.92O_2 \rightarrow CO_2 + 0.375H_2$	−181.85 (exothermic)
r7	$CH_{0.75}O_{0.16} + 0.61O_2 \rightarrow CO + 0.375H_2O$	10.68 (endothermic)
r8	$CH_{0.75}O_{0.16} + 0.42O_2 \rightarrow CO + 0.375H_2$	101.35 (endothermic)

Table 1.
Balanced stoichiometric reactions for Bosjesspruit coal with the addition of methane.

The region bound by the stoichiometric reactions (r1-r8) shaded in light grey, referred to as the stoichiometric or gasification region, represents an important mass balance constraint for any gasification process that uses coal, methane and oxygen as feed. It is in this stoichiometric region that all solid carbon will convert to gas and all methane will reform to gas comprising H_2, CO, CO_2 and H_2O only. Operating out of this region (dark grey regions), by changing feed stoichiometry, will lead to excess amounts of feed not converting in the gasification process and hence be an inefficient conversion process and undesirable.

2.4 Thermally balanced reactions and region

The thermal nature of the reactions (endothermic or exothermic) indicated by the intersections can be used to determine the thermally balanced region (TBR) of operation. The TBR limits the gasification region such that there is no nett heat added or released to the gasifier; the endo- and exothermic reactions are paired thereby resulting in thermally balanced points with a heat of reaction of zero (kJ/mol). To develop a boundary around all the thermally balanced reactions, the extreme reactions, which are said to be linearly independent, are used. Operating within the thermally balanced region is desirable as it increases the thermal efficiency of the plant, eases operation and makes provision for economic savings. Generally, real and practical gasifiers will operate slightly away from the TBR on the "hot" side of the balance in order to use this excess heat to pre-heat the feed and or offset heat losses in the gasifier as explained by [2].

To determine the thermally balanced region of operation, the exothermic reactions (r1 to r6) are balanced with the endothermic reactions r7 and then r8. This results in the 12 reactions in **Table 2**. For example, r1 is balanced with r7 as follows:

$$r_1 + r_7 = \left(\frac{10.68}{35.50}\right)CH_4 + \left(\frac{10.68}{35.50}\right)0.5O_2 \rightarrow \left(\frac{10.68}{35.50}\right)CO + \left(\frac{10.68}{35.50}\right)2H_2 \quad (4)$$

$$+CH_{0.75}O_{0.16} + 0.61O_2 \rightarrow CO + 0.375H_2O$$

$$= CH_{0.75}O_{0.16} + 0.7579O_2 + 0.3007CH_4 \rightarrow 1.3007CO + 0.375H_2O + 0.6014H_2$$

$\therefore$ The heat of reaction for $r_1 + r_7 = \left(\frac{10.68}{35.50}\right)(-35.50) + 10.68 = 0\ kJ/mol$.

No.	Reaction
A	$CH_{0.75}O_{0.16} + 0.7579O_2 + 0.3007CH_4 \rightarrow 1.3007CO + 0.375H_2O + 0.6014H_2$
B	$CH_{0.75}O_{0.16} + 0.641O_2 + 0.0335CH_4 \rightarrow 0.0335CO_2 + CO + 0.375H_2O + 0.067H_2$
C	$\mathbf{CH_{0.75}O_{0.16} + 0.6383O_2 + 0.0206CH_4 \rightarrow 1.0206CO + 0.4161H_2O}$
D	$CH_{0.75}O_{0.16} + 0.6341O_2 + 0.0133CH_4 \rightarrow 0.0133CO_2 + CO + 0.4016H_2O$
E	$\mathbf{CH_{0.75}O_{0.16} + 0.6263O_2 \rightarrow 0.0377CO_2 + 0.9623CO + 0.375H_2O}$
F	$CH_{0.75}O_{0.16} + 0.6248O_2 \rightarrow 0.0554CO_2 + 0.9446CO + 0.3542H_2O + 0.0208H_2$
G	$\mathbf{CH_{0.75}O_{0.16} + 1.8475O_2 + 2.8549CH_4 \rightarrow 3.8549CO + 6.0849H_2}$
H	$\mathbf{CH_{0.75}O_{0.16} + 0.738O_2 + 0.318CH_4 \rightarrow 0.318CO_2 + CO + 1.011H_2}$
I	$CH_{0.75}O_{0.16} + 0.7129O_2 + 0.1952CH_4 \rightarrow 1.1952CO + 0.3905H_2O + 0.375H_2$
J	$CH_{0.75}O_{0.16} + 0.6726O_2 + 0.1263CH_4 \rightarrow 0.1263CO_2 + CO + 0.2526H_2O + 0.375H_2$
K	$CH_{0.75}O_{0.16} + 0.6064O_2 \rightarrow 0.2711CO_2 + 0.7289CO + 0.1017H_2O + 0.2733H_2$
L	$\mathbf{CH_{0.75}O_{0.16} + 0.5989O_2 \rightarrow 0.3579CO_2 + 0.6421CO + 0.375H_2}$

Table 2.
Thermally balanced reactions for Bosjesspruit coal with the addition of methane.

The BE compositions for the thermally balanced equations are calculated and represented graphically by the black region in **Figure 3**. The enclosed area represented by the 5 reactions shows the thermally balanced region of operation – all thermally balanced reaction operate in this area. The area of operation to the right of the region represents exothermic reactions where the gasification syngas product comes out "hotter" than the feed. Also, the area to the left of the region represents endothermic reactions where products come out "colder" than the feed temperature.

Table 2 summarises the 5 thermally balanced reactions in bold (C,E,G,H,L) that form the extreme boundary points and any other thermally balanced reaction (A,B,D,

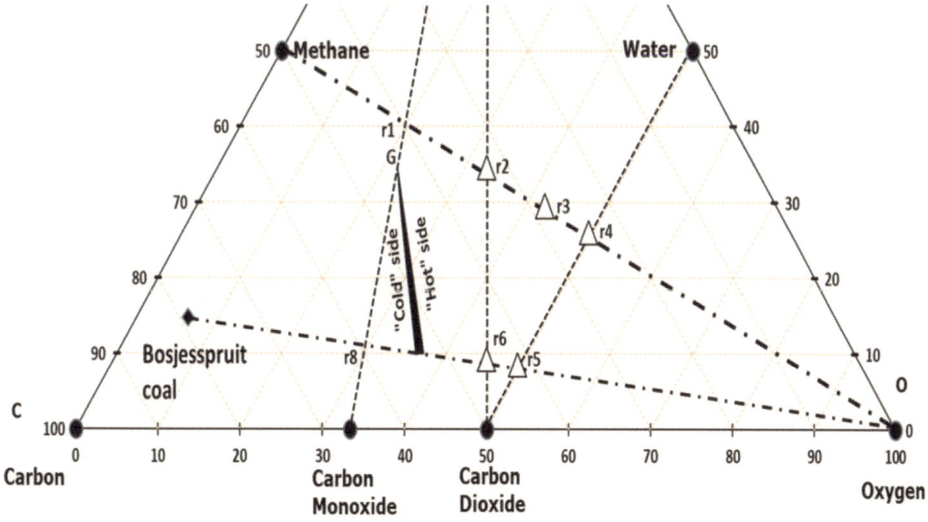

Figure 3.
Thermally balanced region for gasification of Bosjesspruit coal with methane.

F,I,J,K) can be obtained by linear combinations of these 5 reactions. The MATLAB$^{(C)}$ *CONVHULL* function was used to determine the extreme thermally balanced points.

3. Implications of thermally balanced regions

A systematic method of obtaining the stoichiometric reactions and thermally balanced region was developed above from which some very important results are noted. Firstly, the co-gasification of coal and methane is theoretically possible to produce syngas comprising H_2, CO, CO_2 and H_2O at varying compositions. Secondly, these results are consistent and independent of the type of gasifier chosen: fixed bed, fluidized bed or entrained flow, allowing to assess a number of gasification types within a single diagram.

Hence, this allows for the targeted approach to designing co-gasification based systems for either IGCC processes or gas to liquids processes and are discussed below.

3.1 Endothermicity of partial oxidation of Bosjesspruit coal

The reactions representing partial combustion of Bosjesspruit coal r7 and r8 (where syngas produced has significant calorific value) are naturally endothermic. This has a major implication to IGCC where only oxygen is used to obtain syngas with high heating value (HHV). Generally, for IGCC and power applications as seen for good quality coals (northern hemisphere), the partial oxidation leads to syngas rich in H_2 and CO only (no/little CO_2) which is desirable as it represents a clean (no CO_2) gas with high HHV. However, for Bosjesspruit coal, it is not sensible to obtain a H_2-CO only gas as that reaction is endothermic (r8). To obtain the onset of exothermic reactions the addition of oxygen will be required as shown in thermally balanced reaction r8, after which further oxygen will lead to more exothermicity (for higher gasification temperatures) of the system. However, this also means that a lower HHV value gas is obtained which not ideal for IGCC operation – this implies that South African coals of low quality will not be used for IGCC purposes if only oxygen is used. From the analysis in Section 2.4 above it is concluded that the Bosjesspruit coal (low quality) can only be used in IGCC application if methane is available and needs to operate between the thermally balanced reaction G and the partial oxidation of methane (r1) (**Figure 3**). This operation also lies on the H_2-CO line that connects point G and r1which produces clean syngas (no CO_2), is exothermic and high HHV suitable for IGCC operations, regardless of gasifier type.

3.2 Equilibrium considerations & excess steam addition

The thermally balanced reactions (C,E,G,H,L) on the edges of the balanced region (**Figure 3**) are targets for a co-gasification process and are obtained by the precise ratios of coal, methane and oxygen in the feed as well as at high operating temperature (>1700 K) and pressures (>20 bar). However, to obtain these precise outcomes also requires the consideration of thermodynamic equilibrium affecting the distribution of the products in the syngas. It was determined that this can be circumvented by the addition and removal of steam in the system as shown by the following example for thermally balanced reaction H chosen arbitrarily:

$$CH_{0.75}O_{0.16} + 0.738O_2 + 0.318CH_4 + 1.17H_2O \rightarrow 0.318CO_2 + CO + 1.011H_2 + 1.17H_2O \quad (5)$$

At equilibrium (without) steam addition the product distribution would have been 1.22CO, $0.79H_2$, $0.096CO_2$, $0.22H_2O$ (1700 K & 50 bar)– the addition of steam (1.17) and the subsequent condensation of the same amount leads to the reactions as written by H. It is in this context that excess steam is implied for this work.

4. Application to coal-to-liquids

A sensible polygeneration facility, as described earlier [1] where both power generation and liquid fuels/chemicals are made, requires that the co-gasification process produce a syngas that is typically rich in H_2 and CO only. A portion of this syngas would be directed to an IGCC unit for power generation and the rest to a gas-to-liquids process. This was shown above that possible operational regions would be on the line connecting thermally balanced point G and the exothermic partial oxidation of methane (r1) – refer to **Figure 3**. Along this line of operation, the gasification reactions are inherently exothermic and it is preferred to operate closer to point G as the reactions are not highly exothermic (for temperature control in the gasifier) and less CH_4 (more coal) is used overall. However, this also means the syngas composition does not have a high H_2-content for the downstream gas-to-liquids conversion. This section covers some of the strategies employed to obtain a syngas feed composition suitable for methanol production as given by Higman & van der Burgt [10] as an example and the details are provided in **Table 3** below. It is noted that this syngas requires the presence of up to 3.5% CO_2 for the catalyst to operate optimally. While traditional methods would deploy an additional Water-Gas-Shift reactor to correct the H_2:CO ratio, in this work it is not implemented due to the additional equipment cost, raising of additional steam and the CO_2 byproduct created directly by the WGS reaction and indirectly by raising steam.

The thermally balanced point at G represents the maximum H_2 that can be produced from the co-gasification process where the syngas product temperature equals the feed temperature. Generally, a real gasification process will operate just off of this point and into the "hot" exothermic side [2]. The methanol feed, point S, is shown in the ternary diagram, **Figure 4**. The task for the designer is to obtain this feed point starting from point G and requires either the addition of excess hydrogen and or the removal of CO_2 to obtain the final methanol feed composition. Several scenarios are provided and discussed where the choice of additional equipment for WGS was avoided.

4.1 Excess hydrogen from electrolysis

To achieve the methanol feed point (see **Figure 4**) from point G, the addition of H_2 and CO_2 is necessary. Microsoft Excel's solver tool is used to calculate the amount of

Component	mol %
CO_2	3.50
CO	27.59
H_2	67.24

Table 3.
Methanol synthesis feed composition [2].

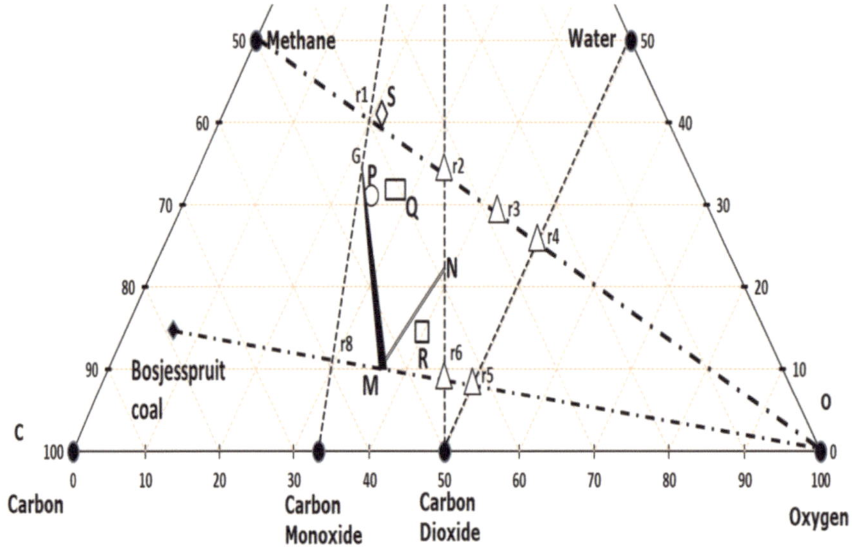

Figure 4.
Target co-gasification points (P, Q) and methanol feed (S).

H_2 and CO_2 required to achieve the correct H_2:CO ratio and CO_2 composition. The target co-gasification reaction (including additional steam for equilibrium consideration – See Section 3.2) is thus determined to be:

$$\begin{aligned} &CH_{0.75}O_{0.16} + 1.4285O_2 + 1.7118CH_4 + 1.6543H_2O \\ &\rightarrow 2.4065CO + 3.7986H_2 + 0.3053CO_2 + 1.6543H_2O \end{aligned} \quad (6)$$

This output is represented by point P on **Figure 4** and represents the syngas operating point from the exit of the co-gasification process. The final methanol feed is obtained by the addition of excess H_2 (such that H_2:CO = 2.44 as required in **Table 3**) obtained from a CO_2-free source such as solar-electrolysis of water. The choice of CO_2-free H_2 introduces the possibility of including renewable resources, to minimise additional CO_2 production, into existing fossil fuel based facilities and in this particular case indicates the minimum amount of renewable H_2 needed for a co-gasification process to exist. For this Bosjesspruit coal the amount of H_2 needed, to obtain the final methanol feed from point P, represents the minimum amount of renewable H_2 needed for co-gasification with methane. Point P may also be implemented for IGCC application as it has a relatively high HHV and is lowest CO_2 in the syngas.

4.2 Obtaining methanol feed by removal of CO_2 from Co-gasification process

Another possibility of obtaining the methanol feed from point G, excluding water electrolysis, is by operating the co-gasification process such that some CO_2 is allowed to be formed allowing the H_2:CO to correct itself internally (no external WGS reactor). This is represented by the point Q on **Figure 4**. From this point, after cleaning the syngas for contaminants (Sulphur, particulates etc) some CO_2 (1.05) is removed to obtain the final methanol feed composition as required. The balanced reaction from the co-gasification process (including additional steam for equilibrium) is thus:

$$CH_{0.75}O_{0.16} + 2.5265O_2 + 2.8549CH_4 + 12H_2O \rightarrow 2.4967CO + 6.0848H_2 + 1.3582CO_2 + 12H_2O \quad (7)$$

4.3 Methanol from traditional gasification of Bosjesspruit coal

The analysis for traditional gasification of Bosjesspruit with steam and oxygen has been done elsewhere [8] and the important thermally balanced reactions are provided in **Table 4** below:

No.	Reaction
M	$CH_{0.75}O_{0.16} + 0.599O_2 \rightarrow 0.642CO + 0.375H_2 + 0.358CO_2$
N	$CH_{0.75}O_{0.16} + 0.544O_2 + 0.752H_2O \rightarrow CO_2 + 1.127H_2$

Table 4.
Thermally balanced reactions for traditional Bosjesspruit gasification [8].

Reaction M and N are used as basis to determine the gasification operation point that is exothermic and lies on the line that connects the final methanol feed and CO_2 point. This is required as the final step requires the removal of CO_2. **Figure 4** shows the line M-N as well as the gasification point R for the coal only system. The Gibbs Free reactor was used on Aspen Plus at 1700 K and 50 bar to determine the equilibrium reaction (amount of H_2O needed for equilibrium). The resulting overall reaction, represented by R (including excess steam for equilibrium), is as follows:

$$CH_{0.75}O_{0.16} + 0.665O_2 + 0.2528H_2O + 6.1H_2O \rightarrow 0.2576CO + 0.6278H_2 + 0.7424CO_2 + 6.1H_2O \quad (8)$$

5. Comparison of processes for poly-generation operation

The three methods to produce syngas for the poly-generation system was described where the H_2:CO ratio required for methanol production was obtained either by co-gasification of coal and methane then adding H_2 from electrolysis using renewable energy (4.1), or from co-gasification with CO_2 removal (4.2) or from the traditional steam-oxygen gasification (4.3) of the same Bosjesspruit coal from a South African mine.

The process reactions are summarised below normalised (per mol of CO) in the syngas produced:

Co-gasification with H_2 addition from Electrolysis

$$0.4156CH_{0.75}O_{0.16} + 0.5936O_2 + 0.7112CH_4 + 0.6875H_2O \rightarrow CO + 2.4371H_2 + 0.1269CO_2 + 0.6875H_2O \quad (9)$$

Co-gasification with CO_2 removal:

$$0.4005CH_{0.75}O_{0.16} + 1.0119O_2 + 1.1435CH_4 + 4.806H_2O \rightarrow CO + 2.4371H_2 + 0.5441CO_2 + 4.806H_2O \quad (10)$$

Process	Excess Steam/CO	Multiple Factor Steam/CO	Excess CO_2/CO	Multiple Factor CO_2/CO	Comment
Co-gasification with H_2 from electrolysis	0.69	1	—		no excess CO_2 but requires electrolysis
Co-gasification with CO_2 removal	4.81	7	0.42	1	requires CO_2 removal – no WGS reactor
Traditional CTL (simultaneous WGS)	23.68	34	2.76	7	requires CO_2 removal – no WGS reactor

Table 5.
H_2O required and CO_2 produced (per Mol of CO) for various processes.

Traditional CTL:

$$\begin{aligned}&3.8819CH_{0.75}O_{0.16} + 2.5815O_2 + 0.9814H_2O + 23.68H_2O\\&\rightarrow CO + 2.437H_2 + 2.8819CO_2 + 23.68H_2O\end{aligned} \tag{11}$$

Table 5 compares the various processes (per mol of CO) against the amount of excess steam required for equilibrium and the excess CO_2 produced.

As seen in **Table 5**, the best case scenario which requires minimal steam and does not produce excess CO_2 is from the co-gasification of coal with methane with the addition of H_2 from renewable-electrolysis. It is noted that the small amount of CO_2 in the methanol feed (or gasification product) is a requirement for optimal catalyst performance and requires the co-gasification process to operate away from the preferred H_2-CO line where equilibrium is favoured. Hence all operations in the ternary diagram that operate away from the H_2-CO line will invariably require excess steam and hence produce excess CO_2 – albeit without a separate WGS reactor.

The traditional CTL process is by far the worst in performance as it produces the most CO_2 and requires the most excess steam. When comparing the two non-electrolysis (for high H_2) processes it is evident that the co-gasification of coal with methane is also superior to the traditional CTL process. Here, the excess steam required for co-gasification process is about 5 times less and up to 7 times less CO_2 is emitted than the traditional CTL process. This is an important result for re-looking at the way traditional CTL is done in South Africa and other developing countries intending to use low grade coal for power generation and or liquids fuel/chemicals production.

6. Conclusions

The co-gasification of coal and methane has been studied from a fundamental understanding of basic mass and energy balances for the purposes of producing syngas for polygeneration facilities where power and liquid fuels are required. Here a South African coal from Bosjesspruit mine is studied in various process routes, namely co-gasification with methane with H_2 addition from water electrolysis using renewable energy, co-gasification with CO_2 removal and the traditional gasification of coal using steam and oxygen. An important result showed that the poor quality of the Bosjesspruit coal requires the co-feeding of methane in the gasification and polygeneration process to produce a syngas rich in H_2-CO ready for IGCC purposes or further treated for liquid fuels/chemicals production. This coal would otherwise be only used for liquid fuels production resulting in high CO_2 emissions and with large requirements for water in the process.

Moreover, a technique of graphical analysis for co-gasification of coal with methane was presented forming the basis for decision making in poly-generation facilities. This allows for the quick screening of coal types as well as strategies required to design co-gasification processes. Lastly, this analysis is independent of the gasification reactor type allowing designers to narrow the options required for their purposes.

Author details

Shaakirah Cassim and Shehzaad Kauchali*
University of the Witwatersrand, Johannesburg, South Africa

*Address all correspondence to: shehzaad.kauchali@wits.ac.za

References

[1] Williams RH. Toward zero emissions from coal in China. Energy for Sustainable Development. 2001;**V**(4): 39-65

[2] Battaerd HAJ, Evans DG. An alternative representation of coal composition data. Fuel. 1979;**58**(2): 105-108

[3] Probstein RF, Hicks RE. Synthetic Fuels. Dover Publications; 2006. pp. 271-276

[4] Cao Y, Gao Z, Jin J, Zhou H, Cohron M, Zhao H, et al. Synthesis gas production with an adjustable H2/CO ratio through the coal gasification process: Effects of coal ranks and methane addition. Energy & Fuels. 2008;**22**:1720-1730

[5] Wu J, Fang Y, Wang Y. Combined coal gasification and methane reforming for production of syngas in a fluidized-bed reactor. Energy & Fuels. 2005;**19**: 512-516

[6] Song X, Guo Z. A new process for synthesis gas by co-gasifying coal and natural gas. Fuel. 2005;**84**: 525-531

[7] Ouyang Z, Guo Z, Duan D, Song D, Wang Z. Experimental study of synthesis gas production by coal and natural gas co-conversion process. Fuel Processing Technology. 2006;**87**:599-604

[8] Kauchali S. Graphical analysis of underground coal gasification: Application of a carbon-hydrogen-oxygen (CHO) diagram. Journal of the Southern African Institute of Mining & Metallurgy 2018;**118**:1067-1078.

[9] Litheko LA. Conceptual Design of Gasification-Based Biorefineries Using the C-H-O Ternary Diagram. Dissertation. UNISA; 2017

[10] Higman C, van der Burgt M. Gasification. 1st. ed. Elsevier; 2003

Chapter 9

Improving Hydrogen Production Yield in Hydrothermal Gasification Processes through Novel Metal Catalysts

Sibel Irmak

Abstract

Catalysts are the most effective and economically feasible way to increase yield of the product(s) in various production processes. The catalysts prepared with innovative approaches could have novel catalytic properties such as increased number of active sites, highly selective to the target product, resistance to deactivation, and extended lifetime. The catalysts with these unique properties could provide significant economic benefits for the production of hydrogen which is currently very expensive. Gasification in hydrothermal conditions has considerable advantages over existing high energy-consuming conversion technologies. Hydrothermal conversion processes take place at mild conditions and wet feed materials such as biomass can be used with no need of drying. However, the absence of practical catalysts in hydrothermal conditions is a main challenge that impedes application of these technologies in large scales. This book chapter focused on the metal catalysts which can be used for hydrothermal gasification processes for high-yielding hydrogen gas production from biomass compounds. The effects of different type of carbon supports, incorporation of heteroatom(s) into catalyst support, different shell structure design, etc., were discussed for hydrogen production in hydrothermal gasification processes.

Keywords: hydrogen, hydrothermal, gasification, biomass, metal catalysts

1. Introduction

Catalysts are the most effective and economically feasible way to increase yield of the product(s) in various production processes. The catalysts prepared with innovative approaches could have novel catalytic properties such as increased number of active sites, highly selective to the target product, resistance to deactivation, and extended lifetime. The precious metal-based catalysts with these properties could provide significant economic benefits for the production of hydrogen which is currently expensive. The hydrothermal gasification technologies (sub- and supercritical water gasification and aqueous-phase reforming) have considerable economic, environmental, and technical advantages over other energy-extensive conversion technologies [1]. These processes are compatible with water-soluble feedstocks such as biomass and gasification reactions that take place at lower temperatures. However, the absence of practical catalysts in hydrothermal conditions is a main challenge that impedes upscaling of these technologies for hydrogen gas production. Increasing demand, limited supply, and undesirable byproducts due to current methods indicate the need for the development of innovative, economically feasible, highly active, and stable catalysts for hydrothermal conversion of biomass-derived compounds to hydrogen in higher yield and richer composition.

2. Catalysts for hydrogen gas production by hydrothermal gasification processes

Raney nickel and platinum-based catalysts are common catalysts that have been used for hydrogen gas production in various processes including hydrothermal gasification methods [2, 3]. Despite Raney nickel catalysts exhibit better activity than precious metals, the use of these catalysts in hydrothermal gasification processes is not the best because of the following reasons: **(a)** Raney nickel deactivates easily in hydrothermal conditions. The studies showed that oxidation of Raney-Ni surface caused the deactivation. The nickel metal particle size in Raney-Ni catalysts became larger, and surface area was significantly decreased during deactivation. To prevent oxidation and deactivation of Raney nickel catalysts, the gasification environment should be reductive as possible to keep Raney-Ni in metallic form and extend its activity [4, 5]. **(b)** Raney nickel catalysts contain significant amounts of hydrogen gas. Because of this, these catalysts can easily ignite when exposed to oxygen. To prevent ignition, Raney nickel catalysts are kept under nonreactive inert environments and stored in water or in a suitable organic solvent such as cyclohexane, ethanol, and dioxane [4]. **(c)** Nickel is not stable under high carbon-containing reaction environments [5]. Therefore, it is hard to keep nickel in its metallic form under organic-rich hydrothermal conditions.

On the other hand, precious metals also show high activities in the reformation of oxygenated hydrocarbons. Since these metals are expensive, they are widely used in the supported form on activated carbon, alumina, titanium dioxide, silica, etc., for recycling [2]. Dispersion of small metal particles on the support with high specific surface area is considered more advantageous. The decrease in the particle size of metals on the support could enhance metal-support interaction and increase the activity of the catalyst for hydrogen production [6]. Large metal particles could cause more severe carbon formation and deposition on the catalyst surface [7].

3. Supported metal catalysts

Dispersion of metals on the support, resistance of metal particles to sintering, and the accessibility of active sites to reactants are highly affected by physical and chemical properties of the catalyst support [8]. Numerous studies showed that catalyst support could significantly affect the catalytic performance of metal catalysts in the various reactions including hydrothermal processes [9, 10]. The catalyst support provides a physical surface for the dispersion of metal particles and affects the catalytic activity [11]. The catalytic activities of the catalysts can be increased by the selection of a novel supportive material for better catalytic action (more active sites) and excellent mechanical strength and high surface area. The supportive materials can be chemically modified to increase surface area and porosity and create specific functional groups on the surface.

3.1 Carbon materials as supportive materials for supported precious metal catalysts

Carbon materials have been recognized as the most active supports for aqueous-phase reforming of biomass-derived compounds for hydrogen-rich gas production [2]. The interaction between active metal component and support plays an important role in the catalytic reactions. Therefore, catalytic properties of supported catalysts depend on the combination of the type of metal and supportive material. The catalytic activity of these catalysts for hydrogen production has been significantly improved with recent research activities; however, they are still not good as Raney nickel catalysts. The activity of the supported precious metal catalysts reduce because of aggregation or poisoning of metal particles in reaction environment or coke deposition on active surfaces of the catalysts [12]. Leaching of the metal particles from support is another problem that causes decay of catalyst activity. Increasing the catalytic activity and stability of the catalysts is a challenge for high-yielding hydrogen gas production.

The porous structure and surface chemistry of activated carbon as a support material highly affect the activity of the catalyst. Porous carbon materials are of interest in many applications due to their high surface area and physicochemical properties. Those carbon materials can be categorized according to their pore sizes as microporous (pore size <2 nm), mesoporous (2 nm $<$ pore size <50 nm), and/or macroporous (pore size >50 nm) [13]. Mesoporous carbons have large surface areas and are rich in oxygen-containing functional groups in the surfaces. They can enhance the affinity between the substrate molecules and the catalyst through abundant functional groups and large uniform pores. Those materials may offer great advantages over other carbon materials owing to their well-controlled pore structures in the mesopores, which might be favorable to the transportation of large molecules such as polysaccharides released from cellulose and hemicellulose structures in the biomass.

The oxygen-containing functional groups on the surface of carbons significantly influenced their performance in catalytic reactions [14]. Many different activating agents such as strong or weak acids and bases or oxidants (H_2SO_4, H_3PO_4, HNO_3, CH_3COOH, NH_3, KOH, H_2O_2, O_3, etc.) can be used in this process to introduce oxygen-containing functional groups on the carbon surface [15, 16]. The abundant oxygen groups in the supports facilitate the access of oxygenated biomass components to the surface of the catalyst; the open pore structure permits the molecules (solubilized biomass components) to effectively diffuse to the active sites and reacts with metal catalyst particles.

Hierarchical porous carbons (HPCs): HPCs are carbon structures with macropores and interconnected meso- and micropores. Porous carbons with large specific surface area and pore volume provide multiple sites for interfacial reaction, molecule retention, and catalytic activity. However, surface area itself is not sufficient for exhibiting exceptional performance because narrow and internal pores may not provide access for some reactants to react.

HPCs structured with micro-, meso-, and macropores can provide multi-active sites in the catalytic conversion processes and better accessibility for the feeds that are composed of different molecular weight organic compounds such as lignocellulosic biomass hydrolysates. The mass transport of different sized biomass compounds can be easily facilitated by the macropores and micro−/mesopores that provide access to the metal catalysts deposited in HPC. Currently, HPC preparation methods are expensive and complex. Synthesis of HPCs by utilization of abundant and widely available waste materials can provide opportunity to develop such materials in a sustainable way for many applications including hydrogen production. Direct carbonization of biomass or any solid organic wastes for the synthesis of HPCs is not applicable because of heterogeneity and uncontrollability in morphology and pore structure of the resulted carbon products. It is almost impossible to produce same carbon materials continuously from direct carbonization of nonuniform solid precursors.

Hierarchical structures with uniform and controllable pore sizes can be prepared by carbonization of different precursors using various templates. Low-density HPCs are prepared from various carbon nanomaterials such as graphene, and the resulted HPCs are aerogel-type carbon materials with moderate surface areas and different sized micro−/mesopore structures [17]. Unfortunately, all these and other existing methods have various drawbacks such as using nonrenewable precursors or templates, long synthesis period (e.g., solvent exchange and supercritical drying in the case of aerogels), involving multiple steps that make the process more costly and time-consuming, etc. [18].

Carbon aerogels are another class of hierarchical porous carbon materials with remarkable physicochemical properties, including low density, large surface area, highly porous structure, and good chemical stability. They can be prepared in different forms (monoliths, beads, powders, or thin films). Different types of carbon aerogels with unique characteristics are suitable for different applications, and some carbon aerogels can exhibit promising catalytic performance in various reactions because of their well-developed and controlled porous structure and high specific surface area.

The most common carbon aerogels are prepared by sol–gel polymerization of resorcinol and formaldehyde mixtures followed with supercritical drying and carbonization that were first prepared by Pekala [19]. In addition to resorcinol-formaldehyde, various monomers can be used to prepare these materials, including melamine-formaldehyde, phenol-formaldehyde, cresol-formaldehyde, phenol-furfural, and some polymers such as polystyrenes and polyurethanes [20–23].

A wide spectrum of different carbon aerogel materials with unique properties can be prepared depending on synthesis and processing conditions. Synthesis conditions of organic aerogel (e.g., resorcinol/formaldehyde ratio, catalysts, and pH), curing and drying methods, and carbonization conditions determine surface area, pore volume, and pore size distribution of final carbon aerogel [24].

Supercritical carbon dioxide and freeze-drying are preferable methods to dry the organic gel while retaining skeletal pore structure during drying stage. During carbonization, dried aerogel is heated under inert atmosphere to obtain carbon-rich structure by removing oxygen and hydrogen functionalities.

Different metal-doped carbon aerogels have been developed including W-, Ru-, Co-, Ni-, Pd-, and Pt-carbon aerogels for various catalytic reaction, and these catalysts can show good activity in hydrogen production processes as well (**Figure 1**).

Graphene-nanostructured carbon materials: Graphene has unique physicochemical properties such as great stability, high surface area, electron mobility, thermal conductivity, and mechanical strength. Graphene is an excellent catalyst support for metal-supported type of catalysts because of its great stability and chemical inertness that are important parameters for increasing lifetime of the catalysts. Graphene has a

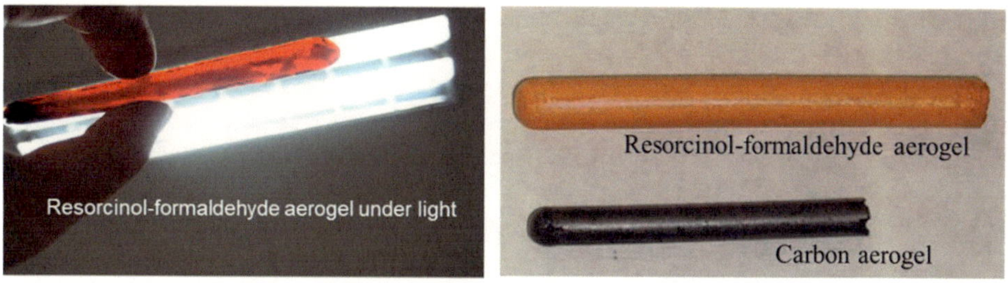

Figure 1.
Resorcinol-formaldehyde aerogel and its carbonized form (carbon aerogel).

high theoretical surface area (approximately 2600 $m^2 g^{-1}$). However, large extent of van der Waals interactions between surfaces lead to graphene sheets to stack on top of each other. Surface area between graphene layers becomes inaccessible depending on the degree of stacking that changes its catalytic activity.

Reactivity of the graphene support can be increased by different approaches. Chemical doping is important approach to tailor the property of graphene that can change its surface reactivity and increase its performance as catalyst support. Introduction heteroatoms, such as nitrogen, boron, phosphorus, or sulfur atoms, into the carbon lattice of graphene by chemical doping can change the electronic properties of graphene [25].

Carbon nanotubes (CNTs) are cylindrical molecules that consist of rolled-up typical graphene sheets (**Figure 2**). Different CNTs (different rolling up direction of graphene layers or single-walled and multi-walled carbon nanotubes) determine the mechanical, electrical, and structural properties of the nanotubes that affect their catalytic activities. Metal particles supported on CNT can poorly be affected from carbon monoxide poisoning than traditional catalyst systems [26], and for this reason CNTs are good candidates to be used as catalyst supports.

A reported study showed that Pt on a single-walled carbon nanotubes catalyst were better catalysts in terms of hydrogen production activity and selectivity than Pt on a multi-walled carbon nanotube for hydrothermal gasification of biomass hydrolysates. Since the biomass hydrolysates tested were composed of large carbohydrate molecules (consisting of two carbohydrate fractions with 69,800 and 25,400 Da), these compounds were unable to enter narrow graphene sheets of multi-walled carbon nanotubes to react with Pt metals deposited inside the graphene layers. On the other hand, when the simplest biomass model compound, glucose, was used as feed

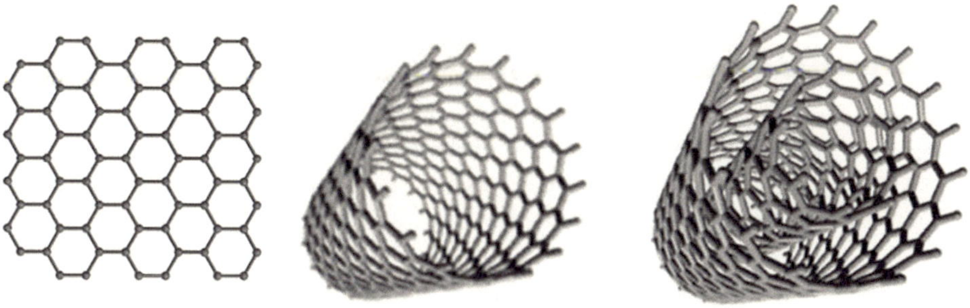

Figure 2.
Chemical structures of graphene, single-walled, and multi-walled carbon nanotubes.

solution, same catalytic activity was observed for both catalysts [27]. These results indicated that the catalytic activity of graphene-based carbon materials is strongly dependent on how the graphene sheet is shaped. Different shaped graphene-based structures with new properties could be promising supportive materials for Pt deposition, and resulting reforming catalysts could exhibit unique properties for hydrogen production (e.g., more active sites and suitable gaps between graphene sheets for the entrance of biomass molecules). The controlled graphene structure permits the small biomass molecules (oligo- and monosaccharides) to effectively diffuse the active sites and react with Pt particles to produce hydrogen gas.

Different graphene nanostructures can be prepared by growing graphene on presynthesized nanostructured metal templates by chemical vapor deposition and then etching away the metal to get a free-standing graphene nanostructure with novel properties to be used as catalyst support in hydrothermal gasification processes [28].

3.2 Heteroatoms-doped carbons as catalyst supports for hydrogen production

The introduction of heteroatom into carbon structure can change the physicochemical and electronic properties of the carbon material and enhance the catalytic functions [29]. Nitrogen atoms can create high positive charge distribution in the nearby C atoms due to its high electron withdrawing ability. Atomic sizes of nitrogen and carbon atoms are similar, and five available valence electrons in nitrogen can lead to the formation of valence bonds with C atoms and covalent bonding between N and C network of carbon results in more stable structure. Dual or multiple heteroatoms (B, N, P, S, F, *etc.*)-doped carbon catalysts enhance the activity of the catalysts [30, 31].

HPCs with heteroatom (e.g., nitrogen and sulfur)-doped carbon network are important carbon-based functional materials and have attracted great attentions because of their excellent properties [32]. Incorporation of nitrogen into the carbon improved the stability of precious metal particles by reducing sintering and leaching of the metals in aqueous-phase furfural hydrogenation reaction [33]. The studies showed that dual or multiple elements-doped carbon catalysts exhibited synergistic effects with enhanced activity in *oxygen reduction reaction* compared with mono heteroatom-doped carbon catalysts [31, 34], and similar effect can be expected for hydrothermal gasification reactions of biomass-derived carbons.

3.3 Increasing activity of the catalysts based on metals deposited on carbon supports

Carbon-supported precious metal (e.g., platinum) catalysts are active catalysts for hydrothermal conversion of biomass-derived compounds to hydrogen [2]. For better hydrogen production yield, Pt particles deposited on the support should be nanosized, uniform, and well dispersed that highly depend on the type of support, deposition method for metals including solvents and other active chemicals used during deposition process, type of metal precursors, reduction method for metal precursors, drying and calcination treatments, etc.

Catalysts can be deactivated during gasification process due to strong adsorption of feed impurities, aggregation or poisoning of metal particles in reaction medium, or coke deposition on active surfaces of the catalysts [12]. These drawbacks can be partially eliminated or lowered by integrating non-precious metals in precious metals containing catalyst system. Catalytic activity of supported Pt catalyst considerably

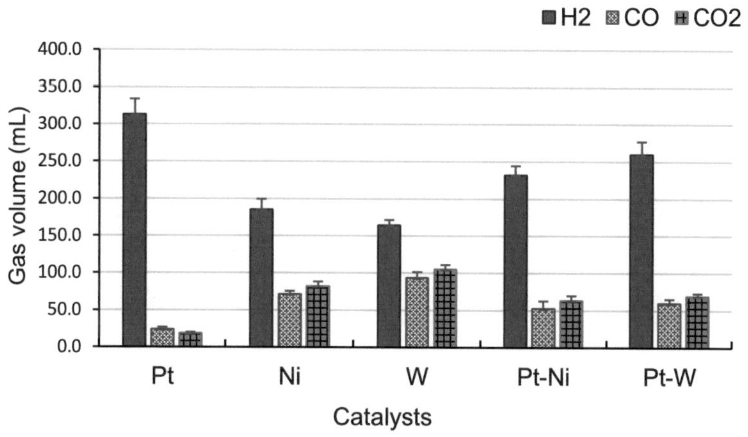

Figure 3.
Comparison of activity of the catalysts prepared with different metals and combination of these metals with Pt for hydrothermal conversion of glucose to hydrogen-rich gas mixture [4].

improved when some certain metals deposited on the support along with Pt. It was reported that the activity of Pt catalysts could be improved by the addition of transition metals that have C-C bond breaking ability (Co, Ni, Fe, Sn, etc.) to a supported Pt catalyst. For instance, the addition of Fe to Pt at a 1:1 Pt:Fe atomic ratio significantly increased hydrogen production yield and selectivity [35]. Addition of a non-expensive metal in precious catalysts can reduce oxidation of precious metal particles by diluting them with non-expensive metals, reduce metal sintering, and favor water-gas shift reaction [36].

Catalytic activity of alumina nanofibers supported Pt catalyst considerably improved when Ni was deposited on the support along with Pt [9]. Kunkes et al. reported that addition of Re to Pt on carbon catalyst enhanced the Pt dispersion and favored water-gas shift reaction [37]. Pt-Co and Pt-Ni catalysts have been reported to be highly active catalysts with high hydrogen selectivity and low coking tendency [7, 38]. Co-deposition of Sn with Pt on to Al_2O_3 support caused dilution of Pt particles with Sn particles that improved activity and selectivity of the catalyst [39].

A recent study showed that performance of Pt-only catalyst could be considerably enhanced by replacing some of Pt particles with different non-expensive metals for hydrothermal conversion of biomass compounds to hydrogen-rich gas mixture. As can be seen in **Figure 3**, the gas mixture produced from hydrothermal gasification of glucose was composed of hydrogen, carbon monoxide, and carbon dioxide. As expected, monometallic Pt catalyst resulted in the highest conversion with highest hydrogen yield as this metal is known to be best catalyst for hydrogen production. Incorporating W and Ni metals to Pt-only catalyst showed promising activity despite replacing half amount of Pt with Ni and W metals [4].

4. Core-shell-type catalysts alternative to supported metal catalysts

Metal core-carbon shell catalysts have exhibited a great potential for various applications, and fabrication of these types of catalysts for specifically hydrogen gas production can be promising. Metal particles are encapsulated in hollow porous carbon shell or heteroatom-doped hollow porous carbon (shell) (**Figure 4**).

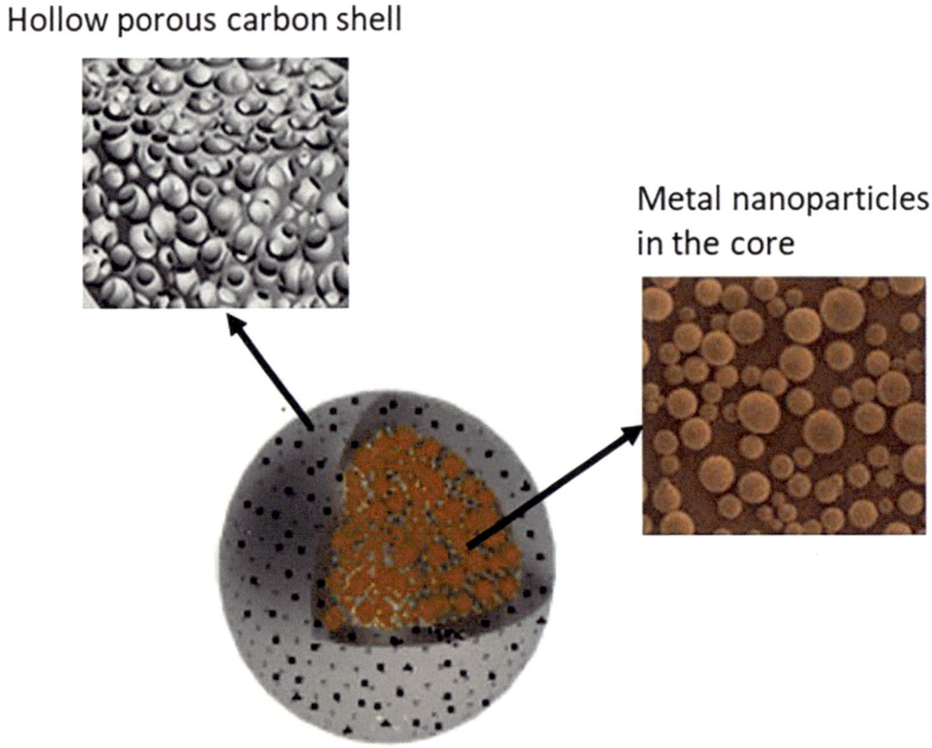

Figure 4.
Core-shell-structured catalysts.

Metal nanoparticles that are made of various combinations of precious metal (e.g., Pt) and/or inexpensive metals (Ni, Sn, Co, and W) in different compositions (e.g., mono and bimetallic) can be used as core materials. It is possible to improve the catalytic activity in this core-shell approach due to increased surface area and the closed interfacial interaction between the core and the shell. Reactants/products can transfer through penetrable shell structure. The shell prevents sintering of the metal nanoparticles in the core. Because of heterogeneous nature, the catalysts can be reused in the process after collecting from aqueous reaction medium by simple separation methods (centrifugation, filtration, etc.).

The encapsulation of metal nanoparticles in a penetrable shell catalyst has recently become a new strategy in the catalyst design area. The encapsulation of precious metal nanoparticles in a stable protective shell can enhance the activity and extent lifetime of the catalyst by maintaining and protecting the size and shape of the precious metal nanoparticles. This core-shell-type catalyst exhibits high stability, catalytic activity, and selectivity in various reactions [40, 41]. For an example, Pt nanoparticles-mesoporous silica core-shell catalyst was reported to have excellent stability for alkene hydrogenation [42].

In the core-shell type of catalysts, the metal nanoparticles are embedded in a protective matrix with channels, which can avoid aggregation or sintering of the metal nanoparticles in the core even at high temperatures and enables transfer of reactants/products through the channels [41]. The metal nanoparticles in the core are mainly responsible for the catalytic activity for a specific reaction. The channels in the shell give access to organic compounds to react with metal particles in the core. This protects metal particles from unwanted reactions in the reaction medium. Ikeda et al. developed such a catalyst design for platinum catalyst for hydrogenation of nitrobenzene in liquid phase [43]. The core-shell catalyst showed very high activity and exhibited almost same activity even after reused. Various porous materials have been used as shell materials to develop such catalysts for a wide range of applications (e.g., mesoporous silica, mesoporous carbon, metal oxides, titanium dioxide, and polymers) [44, 45].

Hollow porous carbon materials have unique properties, such as low density, controllable morphologies, available cavities, high surface areas, tunable porosity, and good chemical stabilities. The encapsulation of metal nanoparticles in a such selectively penetrable shell, hollow porous carbon, can result in novel catalysts for hydrothermal conversion technologies. Hollow porous materials not only have pores in their shell but also contain a hollow core and have higher porosity. Porous carbon shell with hollow structure can enhance the overall activity of the catalysts by increasing the accessibility of the reactants to the active phase. These nanostructures have emerged as an important class of carbon materials in many fields including energy, catalysis, and nanomedicine.

5. Conclusions and future trends

Lack of economically feasible, highly active, and stable catalysts for hydrothermal conversion of biomass-derived compounds to hydrogen is a main challenge that impedes application of these technologies for large-scale hydrogen gas production systems. Choice of appropriate inexpensive metals with right combination of precious metal(s) could reduce catalyst cost and improve the catalytic activity of precious metal-based catalysts. However, the stability of these catalysts in an aqueous processing environment is an important issue that needs to be studied in detail.

Since protection of metal catalyst is a big challenge in hydrothermal condition in which organic-rich solutions/materials are used as feeds, the metal core-carbon shell catalysts could also be a solution. Hollow porous carbon materials can be synthesized in desirable morphologies, sizes, compositions, and pore structures depending on templating strategies used in the preparation steps. The physicochemical and electronic properties of the hollow porous carbon materials can be changed by the introduction of a heteroatom such as nitrogen into carbon shell that enhance the activity and stability of the catalyst.

Graphene-based carbons with controlled graphene nanostructures could show unique properties as catalyst support for metal particles and effectively diffuse the biomass compounds to the active sites and leave agglomeration byproducts and large-sized contaminants in the solution.

Author details

Sibel Irmak
Department of Agricultural and Biological Engineering, Pennsylvania State University, University Park, PA, USA

*Address all correspondence to: sji5137@psu.edu

References

[1] Bamaca Saquic BE, Irmak S, Wilkins M, Smith T. Effect of precursors on graphene supported platinum monometalic catalysts for hydrothermal gasification of biomass compounds to hydrogen. Fuel. 2021;**290**:120079. DOI: 10.1016/j.fuel.2020.120079

[2] Meryemoglu B, Hesenov A, Irmak S, Atanur OM, Erbatur O. Aqueous-phase reforming of biomass using various types of supported precious metal and raney-nickel catalysts for hydrogen production. International Journal of Hydrogen Energy. 2010;**2010**(35):12580-12587. DOI: 10.1016/j.ijhydene.2010.08.046

[3] Cortright RD, Davda RR, Dumesic JA. Hydrogen from catalytic reforming of biomass-derived hydrocarbons in liquid water. Nature. 2002;**418**:964-967. DOI: 10.1038/nature01009

[4] Bamaca Saquic BE, Irmak S, Wilkins M. Enhancement of catalytic performance of graphene supported Pt catalysts by Ni and W for hydrogen gas production by hydrothermal gasification of biomass-derived compounds. Fuel. 2022;**308**:122079. DOI: 10.1016/j.fuel.2021.122079

[5] Nguyen HT, Lu H, Kobayashi E, Ishikawa T, Komiyama M. Raney-nickel catalyst deactivation in supercritical water gasification of ethanol fermentation stillage and its mitigation. Topics in Catalysis. 2014;**57**:1078-1084. DOI: 10.1007/s11244-014-0272-x

[6] Li S, Li M, Zhang C, Wang S, Ma X, Gong J. Steam reforming of ethanol over Ni/ZrO_2 catalysts: Effect of support on product distribution. International Journal of Hydrogen Energy. 2012;**37**:2940-2949. DOI: 10.1016/j.ijhydene.2011.01.009

[7] Iulianelli A, Palma V, Bagnato G, Ruocco C, Huang Y, Veziroglu NT, et al. From bioethanol exploitation to high grade hydrogen generation: Steam reforming promoted by a Co–Pt catalyst in a Pd-based membrane reactor. Renewable Energy. 2018;**119**:834-843. DOI: 10.1016/j.renene.2017.10.050

[8] Feng S, Li W, Shi Q, Li Y, Chen J, Ling Y, et al. Synthesis of nitrogen-doped hollow carbon nanospheres for CO2 capture. Chemical Communications. 2014;**50**:329-331. DOI: 10.1039/C3CC46492J

[9] Tanksale A, Beltramini JN, Dumesic JA, Lu GQ. Effect of Pt and Pd promoter on Ni supported catalysts-A TPR/TPO/TPD and microcalorimetry study. Journal of Catalysis. 2008;**258**:366-377. DOI: 10.1016/j.jcat.2008.06.024

[10] Meryemoglu B, Irmak S, Hesenov A, Erbatur O. Preparation of activated carbon supported Pt catalysts and optimization of their catalytic activities for hydrogen gas production from the hydrothermal treatment of biomass-derived compounds. International Journal of Hydrogen Energy. 2012;**37**:17844-17852. DOI: 10.1016/j.ijhydene.2012.09.024

[11] Kaya B, Irmak S, Hasanoglu A, Erbatur O. Evaluation of various carbon materials supported Pt catalyts for aqueous-phase reforming of lignocellulosic biomass hydrolysate. International Journal of Hydrogen Energy. 2014;**39**:10135-10140. DOI: 10.1016/j.ijhydene.2014.04.180

[12] Ren N, Yang Y-H, Shen J, Zhang Y-H, Xu H-L, Gao Z, et al. Novel, efficient hollow zeolitically microcapsulized noble metal catalysts. Journal of Catalysis.

2007;**251**:182-188. DOI: 10.1016/j.jcat.2007.07.009

[13] Liang C, Li Z, Dai S. Mesoporous carbon materials: Synthesis and modification. Angewandte Chemie International Edition. 2008;**47**:3696-3717. DOI: 10.1002/anie.200702046

[14] Rodriguez-Reinoso F. The role of carbon materials in heterogeneous catalysis. Carbon. 1998;**36**:159-175. DOI: 10.1016/S0008-6223(97)00173-5

[15] Kyotani T, Nakazaki S, Xu WH, Tomita A. Chemical modification of the inner walls of carbon nanotubes by HNO_3 oxidation. Carbon. 2001;**39**:782-785. DOI: 10.1016/S0008-6223(01)00013-6

[16] Chen Z, Hauge RH, Smalley RE. Ozonolysis of functionalized single-walled carbon nanotubes. Journal of Nanotechnology. 2006;**6**:1935-1938. DOI: 10.1166/jnn.2006.336

[17] Xu Y, Lin Z, Zhong X, Huang XQ, Weiss NO, Huang Y, et al. Holey graphene frameworks for highly efficient capacitive energy storage. Nature Communications. 2014;**5**:4554. DOI: 10.1038/ncomms5554

[18] Nishihara H, Kyotani T. Templated nanocarbons for energy storage. Advanced Materials. 2012;**24**:4473-4498. DOI: 10.1002/adma.201201715

[19] Pekala RW. Low density, resorcinol-formaldehyde aerogels. US patent 1989, 4873218

[20] Rasines G, Lavela P, Macias C, Zafra MC, Tirado JL, Parra JB, et al. N-doped monolithic carbon aerogel electrodes with optimized features for the electrosorption of ions. Carbon. 2015;**83**:262-274. DOI: 10.1016/j.carbon.2014.11.015

[21] Zhu YD, Hu HQ, Li WC, Zhang XY. Cresol-formaldehyde based carbon aerogel as electrode material for electrochemical capacitor. Journal of Power Sources. 2006;**162**:738-742. DOI: 10.1016/j.jpowsour.2006.06.049

[22] Wu DC, Fu RM. Synthesis of organic and carbon aerogels from phenol-furfural by two-step polymerization. Microporous and Mesoporous Materials. 2006;**96**:115-120. DOI: 10.1016/j.micromeso.2006.06.022

[23] Ma QL, Cheng HF, Fane AG, Wang R, Zhang H. Recent development of advanced materials with special wettability for selective oil/water separation. Small. 2016;**12**:2186-2202. DOI: 10.1002/smll.201503685

[24] Moreno-Castilla C, Maldonado-Hodar FJ. Carbon aerogels for catalysis applications: An overview. Carbon. 2005;**43**:455-465. DOI: 10.1016/j.carbon.2004.10.022

[25] Fan M, Feng Z, Zhu C, Chen X, Chen C, Yang J, et al. Recent progress in 2D or 3D N-doped graphene synthesis and the characterizations, properties, and modulations of N species. Journal of Materials Science. 2016;**51**:10323-10349. DOI: 10.1007/s10853-016-0250-8

[26] Lin Y, Cui X, Yen C, Wai CM. Platinum/carbon nanotube nanocomposite synthesized in supercritical fluid as electrocatalysts for low-temperature fuel cells. The Journal of Physical Chemistry. B. 2005;**109**:14410-14415. DOI: 10.1021/jp0514675

[27] Kaya B, Irmak S, Hesenov A, Erbatur O, Erkey C. Effect of support materials on supported platinum catalyst prepared using a supercritical fluid deposition technique and their catalytic performance for hydrogen-rich gas production from lignocellulosic biomass.

Bioresource Technology. 2012;**123**:723-726. DOI: 10.1016/j.biortech.2012.07.098

[28] Wilson PM, Mbah GN, Smith TG, Schmidt D, Lai RY, Hofmann T, et al. Three-dimensional periodic graphene nanostructures. Journal of Materials Chemistry. 2014;**C2**:1879-1886. DOI: 10.1039/C3TC32277G

[29] Li D, Li X, Gong J. Catalytic reforming of oxygenates: State of the art and prospects. Chemical Reviews. 2016;**116**:11529-11653. DOI: 10.1021/acs.chemrev.6b00099

[30] Higgins DC, Hoque MA, Hassan F, Choi JY, Kim B, Chen Z. Oxygen reduction on graphene–carbon nanotube composites doped sequentially with nitrogen and sulfur. ACS Catalysis. 2014;**4**:2734-2740. DOI: 10.1021/cs5003806

[31] Sharma PP, Wu J, Yadav RM, Liu M, Wright CJ, Tiwary CS, et al. Nitrogen-doped carbon nanotube arrays for high-efficiency electrochemical reduction of CO2: On the understanding of defects, defect density, and selectivity. Angewandte Chemie, International Edition. 2015;**54**:13701-13705. DOI: 10.1002/anie.201506062

[32] Jang JW, Lee CE, Lyu SC, Lee TJ, Lee C. Structural study of nitrogen-doping effects in bamboo-shaped multiwalled carbon nanotubes. Applied Physics Letters. 2004;**84**:2877-2879. DOI: 10.1063/1.1697624

[33] Huo JJ, Duan P, Pham HN, Chan YJ, Datye AK, Schmidt-Rohr K, et al. Improved hydrothermal stability of Pd nanoparticles on nitrogen-doped carbon supports. Catalysis Science & Technology. 2018;**8**:3548-3561. DOI: 10.1039/C8CY00947C

[34] Liang J, Jiao Y, Jaroniec M, Qiao SZ. Sulfur and nitrogen dual-doped mesoporous graphene electrocatalyst for oxygen reduction with synergistically enhanced performance. Angewandte Chemie (International Ed. in English). 2012;**51**:11496-11500. DOI: 10.1002/anie.201206720

[35] Huber G, Shabaker J, Evans S, Dumesic J. Aqueous-phase reforming of ethylene glycol over supported Pt and Pd bimetallic catalysts. Applied Catalysis B: Environmental. 2006;**62**:226-235. DOI: 10.1016/j.apcatb.2005.07.010

[36] James OO, Maity S, Mesubi MA, Ogunniran KO, Siyanbola TO, Sahu S, et al. Towards reforming technologies for production of hydrogen exclusively from renewable resources. Green Chemistry. 2011;**13**:2272-2284. DOI: 10.1039/C0GC00924E

[37] Kunkes EL, Simonetti DA, Dumesic JA, Pyrz WD, Murillo LE, Chen JGG, et al. The role of rhenium in the conversion of glycerol to synthesis gas over carbon supported platinum-rhenium catalysts. Journal of Catalysis. 2008;**260**:164-177. DOI: 10.1016/j.jcat.2008.09.027

[38] Palma V, Castaldo F, Ciambelli P, Iaquaniello G, Capitani G. On the activity of bimetallic catalysts for ethanol steam reforming. International Journal of Hydrogen Energy. 2013;**38**:6633-6645

[39] Macleod N, Fryer JR, Stirling D, Webb G. Deactivation of bi and multimetallic reforming catalysts: Influence of alloy formation on catalyst activity. Catalysis Today. 1998;**46**:37-54

[40] Tian H, Li X, Zeng L, Gong J. Recent advances on the design of group VIII base-metal catalysts with encapsulated structures. ACS Catalysis. 2015;**5**:4959-4977. DOI: 10.1021/acscatal.5b01221

[41] Kondo T, Morimura T, Tsujimoto T, Aikawa T, Yuasa M. Platinum nanoparticle-embedded porous diamond spherical particles as an active and stable heterogeneous catalyst. Scientific Reports. 2017;7:8651. DOI: 10.1038/s41598-017-08949-0

[42] Lin KJ, Chen LJ, Prasad MR, Cheng CY. Core-shell synthesis of a novel, spherical, mesoporous silica/platinum nanocomposite: Pt/PVP@MCM-41. Advanced Materials. 2004;**16**:1845-1849. DOI: 10.1002/adma.200400349

[43] Ikeda S, Ishino S, Harada T, Okamoto N, Sakata T, Mori H, et al. Ligand-free platinum nanoparticles encapsulated in a hollow porous carbon shell as a highly active heterogeneous hydrogenation catalyst. Angewandte Chemie, International Edition. 2006;**45**:7063-7066. DOI: 10.1002/anie.200602700

[44] Cargnello M, Grzelczak M, Rodriguez-Gonzalez B, Syrgiannis Z, Bakhmutsky K, La Parola V, et al. Multiwalled carbon nanotubes drive the activity of metal@oxide core-shell catalysts in modular nanocomposites. Journal of the American Chemical Society. 2012;**134**:11760-11766. DOI: 10.1021/ja304398b

[45] Chou K-S, Lin M-Y, Wu H-H, Taiwan J. Adsorptive purification of crude glycerol by sewage sludge-derived activated carbon prepared by chemical activation. Institute of Chemical Engineering. 2013;**44**:228. DOI: 10.1016/j.cej.2013.05.120